IRRADIATION FOR FOOD SAFETY AND QUALITY

Proceedings of FAO/IAEA/WHO
International Conference on Ensuring the Safety and Quality
of Food through Radiation Processing

EDITED BY

Paisan Loaharanu

Head, Food and Environmental Protection Section
Joint FAO/IAEA Division of Nuclear Techniques in Food and Agriculture
A-1400 Vienna, Austria

Paul Thomas

Consultant, International Consultative Group on Food Irradiation (ICGFI)
Food and Environmental Protection Section
Joint FAO/IAEA Division of Nuclear Techniques in Food and Agriculture
A-1400 Vienna, Austria

TECHNOMIC PUBLISHING CO., INC.
LANCASTER · BASEL

Irradiation for Food Safety and Quality
a **TECHNOMIC**® publication

Technomic Publishing Company, Inc.
851 New Holland Avenue, Box 3535
Lancaster, Pennsylvania 17604 U.S.A.

Printed in the United States of America
10 9 8 7 6 5 4 3 2 1

Main entry under title:
Irradiation for Food Safety and Quality

Library of Congress Catalog Card No. 00-110712
ISBN No. 1-58716-081-1

HOW TO ORDER THIS BOOK

BY PHONE: 800-233-9936 or 717-291-5609, 8AM–5PM Eastern Time
BY FAX: 717-295-4538
BY MAIL: Order Department
Technomic Publishing Company, Inc.
851 New Holland Avenue, Box 3535
Lancaster, PA 17604, U.S.A.
BY CREDIT CARD: American Express, VISA, MasterCard
BY WWW SITE: http://www.techpub.com

PERMISSION TO PHOTOCOPY—POLICY STATEMENT

Contents

FOREWORD

Significant developments on the acceptance and application of food irradiation as a method to ensure food safety and quality and to facilitate food trade have occurred in recent years. Regulations on food irradiation in many countries either have been or are being harmonized based on the Codex General Standard for Irradiated Foods and relevant recommendations of the International Consultative Group on Food Irradiation (ICGFI). The number of irradiation facilities for treating food is increasing and many more are under construction or being planned. The consumers are getting accurate information and are beginning to appreciate the benefit of irradiated foods. Indeed, irradiated foods are being produced in increasing volume both for food processing industry and retail sales. The potential of irradiation as a method to ensure the hygienic quality of food, especially those of animal origin, as a quarantine treatment of fresh horticultural commodities, and as a substitute for fumigants, is being realized. Irradiation is being considered as a legitimate sanitary and phytosanitary treatment of food and agricultural commodities based on the provisions of the Agreement on the Application of Sanitary and Phytosanitary Measures of the World Trade Organization.

An International Conference on Ensuring the Safety and Quality of Food through Radiation Processing was therefore convened by FAO, IAEA and WHO in Antalya, Turkey, 19-22 October to assess the achievements on food irradiation during the 20th century and its future role as a sanitary and phytosanitary treatment for food and agricultural commodities, current regulatory practices for controlling the application of this process, communication strategies to consumer and the food industry, and to develop action plan for the next century. It was attended by over 150 scientists, regulatory officials and members of the food industry from 48 countries.

The Conference reaffirmed the view that the safety and nutritional adequacy of irradiated food produced under conditions of Good Manufactering Practice is no longer in question, regardless of the absorbed dose.

The widespread and increasing incidence of foodborne illness caused by pathogenic bacteria and parasites and the consequent social and economic impact on human population have brought food safety to the forefront of public health concerns. The Conference agreed that irradiation as a cold pasteurization/decontamination treatment of food, both of animal and plant origin, is an essential step in the Hazard Analysis Critical Control Point (HACCP) based approach, being widely applied or even mandate in many countries, to ensure their safety. There is a great deal of food industry interest in applying irradiation as a sanitary treatment following a number of food-borne disease outbreaks caused by several types of pathogenic bacteria and parasites and massive recalls of ground beef and ready-to-eat food in the USA in the past two years. A consortium of the food industry in the USA with a combined sales of over $2,000 billion per annum has become supportive of the use of irradiation for this purpose, based on its petition submitted to the US Food and Drug Administration in August 1999 to approve irradiation for a number of ready-to-eat food. A number of irradiation facilities available for processing food to ensure their hygienic quality and to meet quarantine requirements have been and are being built in several Asian countries and the USA. Thus, international trade in irradiated food is imminent in the near future.

Sincere appreciation is expressed to Turkish Atomic Energy Authority for hosting the Conference at the beautiful surroundings of Antalya.

SUMMARY OF THE CONFERENCE AND ITS CONCLUSIONS

OBJECTIVES

The specific goals of the Conference were -:

- to assess the past achievements in the practical application of food irradiation processing;
- to identify key issues in further development of food irradiation to ensure food safety, quality and security; facilitate international trade in agricultural commodities; and,
- to develop an agenda for its wider application in the next century.

The Conference was attended by 144 designated participants from 46 Member States including 16 observers from 5 inter-governmental organizations.

The Conference included discussions on the contribution of the technology in ensuring microbial safety of food; enhancing food quality, security and international trade; marketing and consumer acceptance; economics; and regulatory aspects.

BACKGROUND

As the world enters the new Millennium, the provision of safe and nutritionally adequate diets for the world's population will increasingly become a major challenge for governments and the food industry. Many countries at different stages of development are experiencing rapid changes in their health and social environments, where governments are faced with decreasing public sector spending. The strain on limited resources will be further compounded by demographic changes, such as expanding urbanization, the increased dependence on stored and processed foods, and insufficient access to safe water and essential facilities for safe food preparation.

The intensification of food production and the consolidation of the food industries present opportunities for foodborne pathogens to infect large numbers of consumers. In recent years, a number of extremely serious outbreaks of foodborne diseases have occurred on virtually every continent, demonstrating both the public health and social significance of foodborne diseases. In spite of great efforts at the national and international levels, progress in combating foodborne diseases has largely been offset by other global trends, including growing consumer demand for foods of animal origin, longer food distribution networks and many basic changes in the way food is produced, transported, processed, prepared and consumed, Globalization of food trade presents a transnational challenge to food safety control agencies to ensure that imported food is safe consistent with consumer and, in some cases, the environmental protection, but are not unduly restrictive of trade.

To meet these challenges, governments, industry and consumers must join together to assure safe and nutritionally adequate food supplies for present and future generations. This includes the use of all safe and appropriate technologies for improving and extending the availability of food. Seen from this perspective, radiation processing will have an important role to play in ensuring the safety and quality of food in the new Millennium.

The 20th Century has witnessed several advances in the technology of food preservation and processing. These include controlled and modified atmospheres (usually in conjunction with refrigeration), high pressure, high electric field pulse, pulsed light, ohmic heating , oscillating magnetic fields, microwave and extrusion cooking. Some of these technologies are now routinely applied commercially while some others are still in the developmental stages. However, no single technology has such diverse applications as radiation processing. Research and development work in the last five decades have demonstrated that radiation processing can contribute to food safety, food security and trade. Proven practical applications include-:

- destruction of pathogenic bacteria and parasites of public health significance in raw and minimally processed foods;
- microbial decontamination of spices and dried vegetable seasonings;
- insect disinfestation of grains and other stored products;
- inhibition of sprouting in bulb, tuber and root crops;
- shelf-life extension of fruits and vegetables by delaying maturation, ripening and microbial spoilage;
- control of insect pests in fresh fruits and vegetables for quarantine purposes;
- enhancement of the refrigerated shelf-life of meat, poultry, seafood and fresh fruit and vegetables by reduction of spoilage causing microorganisms.

The Conference took place at a time when there is increasing acceptance and application of irradiation as a sanitary and phytosanitary treatment as influenced by-:

- Regulations on food irradiation in several countries and regions either have been or are being harmonized based on the Codex General Standard for Irradiated Foods and relevant recommendations of the International Consultative Group on Food Irradiation (ICGFI);
- The Sanitary and Phytosanitary Agreement of the World Trade Organisation requires that any measures to protect human, animal and plant health must be based on the standards and recommendations of the recognised international authorities. Such authorities include the Codex Alimentarius Commission;
- Irradiation as a method to ensure the hygienic quality of food, especially those of animal origin, as a quarantine treatment of fresh horticultural commodities, and as a substitute for fumigants has been given fresh impetus, especially in the U.S.A., through the regulatory environment, the positive attitude of several major industry and trade organisations and commercial initiatives;

- In 1997 an FAO/IAEA/WHO Study Group on High Dose Irradiation concluded that foods treated with doses greater than 10 kGy can be considered safe and nutritionally adequate when produced under Good Manufacturing Practice. The Codex Alimentarius Commission has already initiated steps to amend the Codex General Standard for Irradiated Foods accordingly;
- Specialised or multi-purpose irradiation facilities are available for treating food in increasing numbers and many more are under construction or are being planned;
- Consumers are receiving accurate information regarding the benefits of food irradiation and as a result are more positive towards its acceptance.

IRRADIATION AND SAFE FOOD

a) Safety of the Process

The Conference agreed that the safety and nutritional adequacy of irradiated foods produced under conditions of Good Manufacturing Practice is no longer in question. The findings of the 1980 JECFI on the safety of any food irradiated up to an overall average dose of 10 kGy were reinforced by an Expert Group convened by the WHO and which reported in 1994. More recently, an FAO/IAEA/WHO Study Group met in 1997 to examine the results of safety studies carried out on foods irradiated with doses higher than 10 kGy. Their 1999 report stated that food irradiated to any dose appropriate to achieve the technological objective is both safe and nutritionally adequate. The Conference endorsed the conclusion of the Study Group that no upper dose limit need be imposed from a food safety standpoint. Similar to other physical food processes, the technical requirements to provide a product that meets technical objectives and has sensory properties acceptable to consumers can determine the upper dose applied to a food.

b) Role of Irradiation as a Sanitary Treatment

The Conference stressed that irradiation is an effective control measure for eliminating pathogenic bacteria and parasites from solid food, especially those eaten raw or minimally processed, of both plant and animal origin without substantial increases in temperature and without causing any significant physical or chemical changes. The need for applying irradiation as a cold pasteurization/decontamination treatment, as an essential step in the Hazard Analysis Critical Control Point (HACCP) based approach, becomes necessary in ensuring their safety. The role of irradiation in combination with other processes and packaging technologies to ensure the hygienic quality of ready-to-eat food, composite food and prepared meals as well as improving shelf-stability of many food products is also likely to increase in the near future.

The widespread and increasing incidence of foodborne illness caused by pathogenic bacteria and parasites and the consequent social and economic impact on the human population have brought food safety to the forefront of public health concerns. Hundreds of millions of people worldwide are affected by diseases caused by contaminated food and the toll in terms of human life and suffering is enormous, particularly among infants and young children, the elderly and other vulnerable groups. At present, food borne disease and the use of irradiation to assist in its control is a focus mainly in developed countries. However, as developing countries become more industrialised, the importance of food borne disease and irradiation will increase further.

Reliable statistics on foodborne diseases are available from very few countries and there is severe under-reporting especially in most developing countries. According to WHO the growing incidence of foodborne diseases already affects between 5-10% of the population each year in industrialized countries. Epidemics of emerging foodborne pathogens such as *Escherichia coli* O157 and *Campylobacter jejuni* in Australia, Japan, Europe and the United states of America, have claimed thousands of victims and caused many deaths. The Conference heard about an improved, active surveillance system (FoodNet) for food borne disease in the USA. Estimates for foodborne illnesses in that country have been revised to about 76 million cases in 1998 amounting to some 30% of the population with about 5,000 deaths.

Outbreaks of foodborne diseases in many industrialized countries are frequently attributed to raw or minimally processed foodstuffs. The young, elderly and immune-suppressed are particularly at risk. Raw foodstuffs including poultry, meat and meat products, seafoods, fruits and vegetables are frequently contaminated with one or several types of food borne bacterial pathogens such as *Salmonella*, *Campylobacter*, *Yersinia*, *Listeria*, *Shigella*, *Vibrio*, *E.coli* O157, and parasites such as protozoa, nematodes and trematodes. These contaminations can result in severe, chronic or fatal health consequences apart from the reduced economic productivity. Government sources in the United States of America estimate the cost of human illness of seven foodborne pathogens to be between US$5.6 to 9.4 billion. The pathogens also pose a liability risk to food companies as shown by massive recalls of some 10,000 tonnes of ground beef in the U.S.A in 1997 because of contamination with *E.coli* O157:H7 and some 7,000 tonnes of ready-to-eat meat contaminated with *Listeria monocytogenes* in late 1998 and early 1999. These recalls have resulted in severe economic losses (estimated to be US$1-3 billions) from destroyed products, liability as well as decreased consumer confidence. Other serious outbreaks of *E.coli* O157:H7 in meats and other minimally processed foods have also occurred in Australia, Japan and Scotland.

c) Role of Irradiation as a Phytosanitary (Quarantine) Treatment of Fresh Fruits and Vegetables

The effectiveness of irradiation as a broad spectrum quarantine treatment of fresh fruits and vegetables is gaining acceptance and application following the endorsement in 1992 by regional plant protection organizations which operate under the framework of the International Plant Protection Convention (IPPC). These organizations include the North American Plant Protection Organization (NAPPO), European Plant Protection Organization (EPPO), Asia and the Pacific Plant Protection Commission (APPPC), Comite de Sanidad Vegetal del Cono Sur (COSAVE), Organism International Regional de Sanidad Agropecuria (OIRSA) etc.

The International Consultative Group on Food Irradiation (ICGFI) has recommended specific treatment schedules for fruit flies of the tephritidae family and other arthropod pests since 1991. The USDA/APHIS issued a policy in 1996 to set a framework for the use of irradiation as a phytosanitary treatment to control 10 species of fruit flies in fresh fruits and vegetables. A large number of countries in Asia and the Pacific agreed in 1999 to implement a harmonized protocol for this purpose. Small scale commercial application of irradiation of tropical fruits to control fruit flies from Hawaii has been successfully implemented in the USA since 1995. A recently concluded international Co-ordinated Research Programme on Use of Irradiation as a Quarantine Treatment of Mites, Nematodes and Insects other than Fruit Flies under the sponsorship of the International Atomic Energy Agency and the Food and Agricultural Organization has shown the potential of irradiation as a phytosanitary treatment for fruits, vegetables, cut flowers and ornamental foliage plants.

The importance of irradiation as an environmentally-friendly phytosanitary treatment is gathering momentum as part of a world-wide trend to reduce chemical treatments of foodstuffs. Ethylene dibromide (EDB) was banned as a fumigant for phytosanitary applications and to control insects in stored food and agricultural commodities in the USA in 1984. Several other countries (e.g., Japan, Germany and other EU countries) also banned EDB. The major fumigant for post-harvest disinfestation is now methyl bromide (MB). The global phase-out of MB in advanced countries by 2005 and in developing countries by 2015 under the Montreal Protocol because of its ozone depleting property, has prompted increasing interest in the use of irradiation as an alternative to MB for insect control in food and agricultural commodities in recent years. The use of MB for quarantine and pre-shipment purposes is exempted from the Montreal Protocol, however.

d) Enhancing Food Security Through Irradiation

Estimates of the amount of post-harvest storage losses of food vary widely, but are known to be unacceptably high (perhaps as high as 30-50%, especially in fruits and vegetables in some developing countries). Insect infestation is the major cause of post-harvest loss in grains, the staple food of most countries. Satisfactory long-term storage of staple crops may be at risk as the traditional fumigant methyl bromide is being phased out and because of the increasing resistance developed by a number of stored product pests to phosphine, the only other major post harvest fumigant used worldwide. Irradiation can contribute significantly to alleviate the post-production losses in staple grain crops caused by insect pests and has several advantages over the traditional post-harvest fumigants as it is a sustainable and environment friendly technology. Irradiation can also reduce post-harvest losses in tuber and bulb crops by inhibiting sprouting in storage under environmental conditions where chemical sprout inhibitors are not effective.

e) Irradiation and International Food Trade

Irradiation of food has considerable potential to increase international trade in agricultural commodities. This presents an opportunity particularly for developing countries as they seek to improve their economies through trade with markets in developed countries. The ability of irradiation to reduce food borne pathogens, to meet quarantine requirements and to extend shelf (or transport) life can all assist in increasing trade.

For example, health authorities in many countries have imposed stricter hygienic standards in food trade. Such standards may require zero tolerance of pathogens such as

Salmonella and *Vibrio cholera* in imported food products such as seafoods. US authorities have already introduced zero tolerance for *Listeria monocytogenes* in ready to eat food in trade and reclassified *E.coli* O157:H7 as an adulterant of raw ground beef and other non-intact meat. Following the ban on ethylene oxide in the EU and Japan, irradiation has provided an effective alternate decontamination process for spices and dried seasonings. In 1998 some 70,000 tons of these products were irradiated worldwide and this is expected to increase in volume in the coming years. In addition, irradiation provides the most versatile treatment for fresh horticultural commodities to overcome quarantine barriers.

REGULATORY ISSUES

The Conference agreed that national regulations -:

- had no need to regulate for maximum dose limits from a toxicological and nutritional perspective provided that Good Manufacturing and Irradiation Practices are maintained;
- should concentrate not on dose limits per se but on the production of microbiologically safe products that meet the stated technical purpose; this may require the imposition of specific conditions of dose or storage in certain situations, for example for phytosanitary uses or where there could be a risk of botulism;
- should provide appropriate flexibility for processors;
- should be in conformity with Codex and take into account the implications of the WTO Agreements.

The Codex General Standard on Food Irradiation recognizes that irradiation is a food process and should be regulated in the same way as other physical processes of foods. However, most countries still regulate on the basis of a case-by-case assessment and as if the treatment is a chemical additive. This is a major cause of variability (non-harmonisation) between national regulations and is a potential barrier to trade.

The Conference was informed of considerable progress towards harmonised regulations in a number of countries and regions, particularly ASEAN. These harmonised regulations are based on the ICGFI Guidelines for the Authorisation of Food Irradiation Generally or by Classes of Food that is itself based on the Codex General Standard. The Harmonised Regulation for Asia and the Pacific is noteworthy for providing only Advisory Technological Dose Limits, not mandatory limits. The Conference also noted that under the WTO Agreements, especially the Sanitary and Phytosanitary (SPS) Agreement, governments which have import regulations stricter than recognized international standards, guidelines and recommendations, may be requested to furnish justifications based on scientific grounds to the WTO. The recognised international standards etc. are those of the Codex Alimentarius Commission, the International Plant Protection Convention and the International Office of Epizootics. Thus, non-tariff barriers to foods by importing countries solely because of an irradiation treatment are no longer justified and may be subject to challenge under the WTO procedures.

FACILITIES AND ECONOMICS

Current trends indicate that both isotopic and machine sources will be used to process food products. There is a wide variety of irradiator designs with varying throughput capacities and capital and operating costs. New developments include -:

- the design of the horizontal Cobalt-60 source concept for efficient processing of meat and poultry products;
- electron accelerators producing energies up to 10 MeV with beam up to 200 kW for treatment of thin products with high throughput rates;
- the availability of high-power electron beam machines permitting the production of X-rays at dose rates and with throughput comparable to radionuclide sources; and,
- low energy electron irradiation for surface treatment of foods for microbial decontamination of grains and seeds; a prototype irradiator is under construction in Japan.

The economics of isotopic and machine sources have been shown to be comparable for specific processing requirements. New commercial scale facilities available for treatment of foods are under construction or about to be commissioned in Brazil, India, Thailand and U.S.A. An electron beam facility for commercial scale food irradiation, being constructed in Sioux City, Iowa, USA, will be able to pasteurize up to 100,000 tonnes of chicken, ground beef or other products per year. The Conference heard that the leading US producer of chicken had entered into an agreement to utilise the EB facility and that another major food consortium were ready to irradiate other meats and meat products. In addition, sales of irradiated chicken treated by a cobalt-60 facility are increasing to both the retail and food service industries due to a successful local educational campaign and collaboration with the State Department of Health in Florida. A nationwide chain has committed to marketing irradiated beef as soon as regulations permit its sale. The availability of irradiation facilities and radiation sources to meet the increasing demand of the food industry on a global scale in the foreseeable future was assured by the suppliers.

CONSUMER ACCEPTANCE AND INDUSTRY INTEREST

Market experience in several countries show that consumers are willing to purchase irradiated food whenever it has been available once they understand the benefit. Numerous consumer acceptance studies and market tests worldwide indicate that majority of the consumers are ignorant about irradiated foods and acceptance increases when consumers are provided with information about the safety and benefit of the process. Consumer information and awareness have shown an upward trend in the recent times through the efforts made by National Governments, food and health professionals, ICGFI, and a responsible media.

Experience in the USA and several Asian countries has indicated that acceptance will increase with increased consumer education and with public endorsements by health professionals. A 1999 survey in USA revealed that a majority of supermarket shoppers are "very/somewhat likely" to purchase food products like strawberries, poultry, pork or beef if they had been irradiated to kill germs and keep it safe.

Despite this evidence, there is still reluctance within the food industry in many countries to adopt the technology. This is due to several reasons that include a perceived consumer resistance, or an unwillingness either to upset the status quo or to be the first to promote a technology that is often regarded as controversial. The recent initiative from the Coalition on Food Irradiation, represented by major food industry and trade associations, co-ordinated by the National Food Processors Association in the USA in petitioning to the Food and Drug Administration (FDA) for approval of irradiation for a variety of fresh, minimally processed and ready to eat foods of both plant and animal origin points to the changing industry interest in food irradiation processing. The Conference agreed that this change needs to be accelerated by the provision of more and better prepared information to leading industry executives.

As the commercial use increases, further research will be required to assist industry address specific issues for particular products such as the phytotoxic effects of radiation on fresh, cut vegetables and the suitability of packaging materials.

CONCLUSIONS

- The Conference agreed that the safety and nutritional adequacy of foods irradiated below and above 10 kGy and produced under Good Manufacturing Practice is now well established.

- Food irradiation is beginning to play an important role in contributing to improved food safety and security and to increased trade as a proven sanitary, phytosanitary and preservation method. This role will be even more important in the new Millennium in view of the increasing awareness of the risk of food-borne pathogens and global trade in food commodities.

- The rapid increase in the global population from 6 billion in 1999 to an expected 9 billion by 2040, a significant proportion of whom will be immuno-compromised, will require a new food safety and security strategy to meet the demand on the food supply of this population increase. As governments have the responsibility to ensure the safety and nutrition of food supplies to their population, it is of critical importance that they adhere to international agreements and standards to improve food availability, safety and trade based on the provisions of the SPS and TBT Agreements of the WTO especially relevant Codex Standards and recommendations of the IPPC.

- Major food importing countries (EU, Japan, USA) are keys to the economy of developing countries which need to export their food supplies to generate foreign exchange. Based on the Codex General Standard for Irradiated Foods and the endorsement of irradiation as a phytosanitary treatment of fresh horticultural commodities by regional plant protection organizations which operate within the framework of the IPPC, trade barriers against food treated by irradiation, still imposed by some major importing countries, are no longer justified and may be subjected to challenge. The Conference urged all governments, especially those from major importing countries, to ensure that their regulations are based on the principle of the Codex General Standard for Irradiated Foods and the relevant recommendations of the International Consultative Group on Food Irradiation, to overcome trade barriers and to encourage the acceptance of food irradiation alongside other food technologies at the earliest opportunity.

- The Conference also considered a number of specific issues, and reached the following recommendations by consensus.

- Food safety regulations should concentrate on production of microbiologically safe products that meet the stated technical purpose rather than on dose limits, and provide appropriate flexibility for processors to provide a quality product. Irradiation should be considered as an integral component of HACCP to ensure microbiological safety of solid foods, especially those eaten raw or minimally processed, equivalent to thermal pasteurization of liquid foods, and to prevent cross contamination during food preparation;

- Regulations should be in conformity with the Standards, Guidelines and Recommendations of the international authorities recognised by the SPS Agreement of the WTO, and in particular with the Codex General Standard for Irradiated Foods and the recommendations of the IPPC;

- Irradiation is established as a versatile, environmentally-friendly treatment of foods for sanitary, phytosanitary and shelf-life extension purposes that can contribute to better food safety, phytosanitary and food security, to reduced use of chemical treatments and to increased trade;

- Market trials indicate that in many countries a majority of consumers are willing to purchase irradiated products; there is, however, a significant barrier to adoption of the technology from the food industry that either perceives significant consumer resistance or has concerns about being seen as the leader in a technology often perceived as controversial;

- Scientists and public health officials have a responsibility to ensure that factual information is continuously generated and presented to key government officials, media and consumer organisations etc., and to influential decision makers in the food trade;

- As specific commercial applications are developed, there will be a need for further research to adapt the treatment to be efficient and effective; examples requiring further research may include the radiation tolerance of irradiation on fresh, cut vegetables and on packaging materials;

- Wherever feasible, new market trials of irradiated foods should be carried out in regions where there is a well established database on existing foodborne disease in order to provide evidence for the effect of irradiation on disease incidence.

Achievements in Food Irradiation during the 20th Century

J. F. DIEHL
Federal Research Center for Nutrition
Karlsruhe, Germany

ABSTRACT

Developments in food irradiation during the last hundred years can be roughly divided into three periods: **First half of the century:** Pioneering studies on the use of ionizing radiaton to destroy food spoiling microorganisms and insects. Lack of suitable radiation sources prevents commercial application. **1950 to 1970:** Technological studies with the aid of newly available radioisotope sources (Co-60, Cs- 137), electron accelerators (Van de Graaff, linear accelerator and others), and - to a smaller degree - X-rays. Search for optimal conditions of irradiation (radiation dose, packaging, atmosphere and temperature during irradiation) for various purposes (sterilization, pasteurization, insect disinfestation, sprout inhibition of potatoes, onions etc.). Hesitation to use the process commercially, caused by doubts about the health safety of consuming irradiated food. **1970 to 1999:** Studies on the wholesomeness of irradiated food, international harmonization of legal regulations, slowly growing commercial application of the process. The years indicated should not be considered as sharp dividing lines. Wholesomeness research was carried out already before 1970 and some technological studies are still going on. A decisive event in the history of food irradiation was the establishment of the International Project in the Field of Food Irradiation (IFIP) in 1970, with the specific aim of sponsoring a worldwide research program on the health safety of irradiated food. Under the sponsorship of FAO, IAEA and OECD, 19 countries joined their resources, with this number later growing to 24. WHO became associated with the International Project in an advisory capacity. The research program included long-term animal feeding studies, short-term screening tests, and the study of chemical changes in foodstuffs irradiated with a dose of up to 10 kGy.

The results obtained in the framework of the International Project and in national testing programs were evaluated by the Joint FAO/IAEA/WHO Expert Committee on the Wholesomeness of Irradiated Food (JECFI) in 1976 and 1980. In 1980 this Committee concluded that the irradiation of any food commodity up to an overall average dose of 10 kGy presents no toxicological hazard and no special nutritional or microbiological problems. On this basis the Codex Alimentarius Commission adopted in 1983 a General Standard for Irradiated Foods and a Recommended International Code of Practice for the Operation of Radiation Facilities. National governments and international agencies which had participated in IFIP felt that the international platform for exchange of information on food irradiation provided by the Project since 1970 has been very useful and should be maintained. As a result of these considerations the International Consultative Group on Food Irradiation (ICGFI) was created in 1983, which is now supported by 46 governments.

In 1997 an FAO/IAEA/WHO Study Group on High Dose Irradiation examined the results of safety studies carried out on food irradiated with doses higher than 10 kGy. Recognizing that in practice the doses applied to eliminate biological hazard would be below those doses that might compromise sensory quality, the **Study Group concluded that no** upper dose limit need be imposed. Accordingly, irradiated foods were deemed wholesome throughout the technologically useful dose range from below 10 kGy to envisioned doses above 10 kGy.

In spite of reassuring research findings and contrary to evaluations by groups of experts, opposition to the practical use of food irradiation is still widespread in several countries. As stated by WHO's Director General: "Decades of study and practical application have fostered increasing confidence in the ability of food irradiation to protect and preserve food and thereby to safeguard health. WHO is concerned that the unwarranted rejection of this process, often based on lack of understanding of what food irradiation entails, may hamper its use in those countries which may benefit most."

Public opinion about food irradiation was generally positive during the growth period of nuclear technology in the 1950s and 60s. With the advance of the antinuclear movement since the 1970s the climate of opinion has changed. Much more than other modern methods of food processing (microwave heating, extrusion cooking etc.), food irradiation has to overcome barriers created by prejudice, misleading information, restrictive legal and regulatory measures and a resulting reluctance of food manufacturers and food trade to make use of the new technology.

Concern about public health problems created by the presence of pathogenic microorganisms in food (Salmonella, Listeria, Campylobacter and others) and the recognition of food irradiation as an effective tool to combat these problems have helped to overcome the barriers. Irradiation of spices to improve their hygienic status is now widely practiced in many countries. The undisputed success of this application will pave the way for other irradiated products.

INTRODUCTION

The story of food irradiation in this century is not an easy one to tell. Many disciplines have been involved: food science and engineering, radiation chemistry and physics, nutrition and microbiology, toxicology and economics. Even sociology should be mentioned, because many studies on the acceptance of irradiated food by consumers were carried out in recent years. Jurisprudence was required to remove legal obstacles to the practical application of this new technology. Research and development work was carried out all over the world, and many countries had national programs on food irradiation. International agencies, such as FAO, IAEA and WHO played important roles in coordinating and supporting these efforts. This complexity makes food irradiation a particularly interesting subject. Only a small fraction of the multifaceted work and of the achievements resulting from it can be mentioned here. Several books and review articles will be quoted, giving the reader the opportunity to dig deeper. Providing a historical oversight, I will divide the century into three periods: the first fifty years (pioneering studies), the period from 1950 to 1970 (technological studies with the aid of newly available radiation sources), and lastly the time from 1970 to 1999 (studies on the wholesomeness of irradiated food, international harmonization of legal regulations, growing commercial application). There is considerable overlap between these periods; wholesomeness research was carried out already before 1970, and technological studies continued after 1970.

The first half of the century

The German physicist W. C. Roentgen discovered X-rays in 1895 and one year later the French physicist A. H. Becquerel observed radiation emitted from uranium. His assistant, Polish chemist Marie Curie and her French husband Pierre, soon after discovered the radioactivity associated with thorium, polonium and radium. It did not take long for enterprising inventors to find practical applications for these radiations. British Patent no. 1609 was issued in 1905 to J. Appleby and A. J. Banks for their invention "to bring about an improvement in the condition of foodstuffs" and in "their general keeping quality". They proposed the treatment of food, especially cereals, with alpha, beta or gamma rays from radium or other radioactive substances, stressing "the exceptionally marked advantage of an entire absence of the direct use or employment of chemical compounds." A U.S. patent for an "Apparatus for Preserving Organic Materials by the Use of X-rays" was granted to D. C. Gillett in 1918. B. Schwartz of the U.S.Department of Agriculture suggested the use of X-rays for inactivating trichinae in pork in 1921. A French patent was granted in 1930 to the German engineer O. Wüst for an invention described by the words: "Foods of all kinds which are packaged in containers are submitted to the action of hard X-rays to kill all bacteria". None of these proposals led to a practical application, simply because the radiation sources available at that time, X-ray machines or radioactive isotopes, were not powerful enough to treat food in commercial quantities.New interest was stimulated in 1947 by the invention of a pulsed electron accelerator, the Capacitron, by A. Brasch and W. Huber, founders of Electronized Chemicals Corporation in Brooklyn, New York. They reported that meats and some other foodstuffs could be sterilized by high-energy electron pulses, that undesirable radiation effects (off-flavors) could be avoided by irradiation in the absence of oxygen and at low temperatures, and that irradiation "will not materially increase the final price of the treated product". At about the same time, J. G. Trump and R. J. van de Graaff of the Massachusetts Institute of Technology, who had developed another type of electron accelerator, also studied the effects of irradiation on food and other biological materials. The foundations of food irradiation research had been laid when B. E. Proctor and S. A. Goldblith of MIT's Department of Food Technology reviewed these early studies in 1951 [1]. More details and literature references on this early period of food irradiation research can be found in Josephson's review [2] and in my book [3].

The period 1950 to 1970

Research and development work in this period was primarily aimed at finding optimal conditions of irradiation (radiation dose, ripeness of produce, packaging, atmosphere and temperature during irradiation) for various purposes (sterilization, pasteurization, insect disinfestation, sprout inhibition of potatoes, onions etc.). In 1950 the U.S. Atomic Energy Commission (USAEC) had initiated a coordinated research program on the use of ionizing radiation for food preservation. Cobalt-60 sources and electron accelerators were made available for this purpose. The U.S.Army supported research on both low- and high-dose food irradiation from 1953 to 1960. After 1960 the Army concentrated its support on high-dose applications, specifically on the development of radiation-sterilized meat products to substitute for canned or frozen military rations. During 1961-62 a large food irradiation laboratory was constructed at the U.S.Army Natick Laboratories in Natick, Massachusetts, which soon became an internationally recognized center of food irradiation research. Reports from the United States about successful experiments on irradiation of food stimulated similar efforts in other countries. In the United Kingdom investigations on the effects of ionizing radiation on food began in 1950 at the Low Temperature Research Station at Cambridge and somewhat later at the Wantage Research Laboratories of the Atomic Energy Research Establishment [4]. By the mid- or late 1950s national research programs on food irradiation were also underway in Belgium, Canada, France, The Netherlands, Poland, the Soviet Union, and the Federal Republic of Germany. The first commercial use of food irradiation occurred in 1957 in Germany, when a spice manufacturer in Stuttgart began to improve the hygienic quality of his products by irradiating them with electrons, using a van de Graaff generator. The machine had to be dismanteled in 1959 when a new food law prohibited the treatment of food with ionizing radiation, and the company turned to fumigation with ethylene oxide instead. In Canada irradiation of potatoes for inhibition of sprouting was allowed in 1960 and Newfield Products Ltd. began irradiating potatoes at Mont St. Hilaire near Montreal, in September 1965. The plant used a cobalt-60 source and was designed to process some 15,000 t of potatoes a year for inhibition of sprouting during long-term storage. It closed after only one season, when the company ran into financial difficulties. In Turkey, at the harbor of Iskenderun, an irradiator for insect disinfestation of grain was built in 1965 with support from IAEA and with funds from the United Nations Development Program. It never received the required operating license from the Turkish Government and finally had to be dismantled in 1968. In spite of these setbacks, interest in food irradiation grew worldwide. At the first International Symposium on Food Irradiation, held in Karlsruhe, Germany, in 1966, representatives from 28 countries reviewed the progress made in national research programs [5]. However, health authorities in these countries still hesitated to grant permissions for marketing irradiated food. At that time only three countries, Canada, the United States, and the Soviet Union, had given clearance for human consumption of a total of five irradiated foodstuffs, all treated with low radiation doses. The food industry had not made use of the permissions; irradiated food was still not marketed anywhere.

The years from 1970 to the present:

Questions about the safety for human consumption of irradiated food were still hotly debated and this was recognized as the major obstacle to commercial utilization of the new process. As a result of this recognition the International Project in the Field of Food Irradiation (IFIP) was created in 1970, a decisive event in the history of food irradiation. The project had the specific aim of carrying out a worldwide research program on the health safety of irradiated food. Under the sponsorship of FAO, IAEA and OECD, 19 countries joined their resources, with this number later growing to 24. Office and laboratory facilities for the project director and his staff were provided at the Federal Research Center in Karlsruhe, and I had the honor and pleasure of serving as the chairman of the Project's Scientific Program Committee. WHO became associated with IFIP in an advisory capacity. The research program included long-term animal feeding studies, short-term screening tests, and the study of chemical changes in foodstuffs irradiated with a dose of up to 10 kGy [3]. The decision to limit the studies to the dose range of up to 10 kGy was based on two considerations: First, most of the commercially interesting applications of the process can be carried out with doses of up to 10 kGy, and second, very extensive toxicological studies with chicken meat irradiated at a sterilizing dose (58 kGy) were underway in the United States at that time. The results obtained in the framework of the International Project and in national testing programs were evaluated by the Joint FAO/IAEA/WHO Expert Committee on the Wholesomeness of Irradiated Food (JECFI) in 1976 and 1980. In 1980 this committee concluded that the irradiation of any food commodity up to an overall average dose of 10 kGy presents no toxicological hazard and no special nutritional or microbiological problems [6,7]. This was a real breakthrough. The International Project had successfully completed its task of examining the wholesomeness of food irradiated at a dose of up to 10 kGy and was terminated. Its results have been documented in more than 60 Technical Reports and in several books, such as [8]. National governments and international agencies which had participated in IFIP felt that the international platform for exchange of information on food irradiation provided by the Project since 1970 had been very useful and should be maintained. As a result of these considerations the International Consultative Group on Food Irradiation (ICGFI) was created in 1984, which is now supported by 46 member countries. The major objectives of ICGFI are to evaluate global developments in the field of food irradiation. ICGFI provides publications on the safety of irradiated food, the effectiveness of food irradiation, commercialization of the process, legislative aspects, control of irradiation facilities, acceptance of and information on food irradiation. ICGFI has also organized training courses for operators, plant managers and technical supervisors of food irradiation facilities as well as for food inspectors and control officials, seminars for journalists, and workshops on technical aspects of food irradiation. Anyone who has access to the Internet may visit the ICGFI Web Page at <www.iaea.org/icgfi>, which contains a lot of information and "hyper-links" on food irradiation. On the basis of JECFI's landmark decision of 1980, the Codex Alimentarius Commission adopted in 1983 a General Standard for Irradiated Foods and a Recommended International Code of Practice for the Operation of Radiation Facilities. An International Conference on the Acceptance, Control of and Trade in Irradiated Food, held in Geneva, Switzerland in 1988, provided an endorsement of the process and its control procedures [9]. The World Health Organization encourages the use of the process, which it described as "a technique for preserving and improving the safety of food" [10]. Nevertheless, doubts about the wholesomeness of irradiated food persisted. Therefore, and at the request of one of its member states, WHO asked another group of experts to review and evaluate the results of scientific studies carried out after 1980, together with the older studies which had already been considered by previous international and national expert committees. The report of this expert group concluded that "food irradiation is a thoroughly tested food technology. Safety studies have so far shown no deleterious effects. Irradiation will help to ensure a safer and more plentiful food supply by extending shelf-life and by inactivating pests and pathogens. As long as requirements for good manufacturing practice are implemented, food irradiation is safe and effective. Possible risks resulting from disregard of good manufacturing practice are not basically different from those resulting from abuses of other processing methods, such as canning, freezing and pasteurization" [11].

In 1997, an FAO/IAEA/WHO Study Group on High Dose Irradiation examined the results of safety studies carried out on food irradiated with doses higher than 10 kGy. Recognizing that in practice the doses applied to eliminate biological hazards would be below those doses that might compromise sensory quality, the study group concluded that no upper dose limit need be imposed. Accordingly, irradiated foods were deemed wholesome throughout the technologically useful dose range from below 10 kGy to envisioned doses above 10 kGy [12]. In spite of the reassuring research findings and contrary to the endorsement of the process by WHO, FAO and numerous scientific societies, opposition to the practical use of food irradiation is still widespread in several countries. Public opinion about food irradiation was generally positive during the growth period of nuclear technology in the 1950s and 60s. With the advance of the antinuclear movement since the 1970s the climate of opinion has changed. Much more than other modern methods of food processing (microwave heating, extrusion cooking etc.), food irradiation has to overcome barriers created by prejudice, misleading information, highly restrictive legal and regulatory measures and a resulting reluctance of food manufacturers and food trade to make use of the new technology. The way in which food irradiation was systematically downgraded and misrepresented in the European Parliament has been documented [13]. The arguments made against irradiation of food are much the same as those expressed at the beginning of this century against pasteurization of milk, as described by Satin [14]. Concern about public health problems created by the presence of pathogenic microorganisms in food (Salmonella, Listeria, Campylobacter and others) and the recognition of food irradiation as an effective tool to combat these problems have helped to overcome the barriers. Irradiation of spices to improve their hygienic status is now widely practiced in many countries. The tonnage of spices and dried seasonings irradiated in those countries has multiplied in recent years, as demonstrated in Fig.1. The food industry in the United States, long hesitant to make use of the process, is now actively pursuing the possibilities of irradiating products other than spices. According to a news release of 23 August 1999, the Food Irradiation Coalition, an alliance of food industry trade associations, health organizations, academic and consumer groups, has asked the U.S. Food and Drug Administration to allow irradiation of ready-to-eat meat and poultry products and fruit and vegetable products for the purpose of eliminating illness-causing microorganisms. The cosponsoring organizations of the Coalition include the National Food Processors Association, the American Association of Meat Processors, the American Bakers Association, the American Chicken Council, the Pacific Seafood Processors Association and several other trade groups. Other cosponsors include the American Society of Microbiology and the Association of Food and Drug Officials. Never before has food irradiation had such massive support in any country. On this basis a growing use of the process can be safely predicted for the coming century - but this is the topic of the following chapter.

Bibliography

1. Proctor, B. E. and Goldblith, S. A. 1951. "Food processing with ionizing radiation". Food Technol. 5: 376-380.
2. Josephson, E. S. 1983. "An historical review of food irradiation". J. Food Safety 5: 161-190.
3. Diehl, J. F. 1995. Safety of Irradiated Foods, 2nd edit. New York: Marcel Dekker, Inc.
4. HMSO, 1964. Report of the Working Party on Irradiation of Food. London: Her Majesty's Stationery Office.
5. IAEA, 1966. Food Irradiation. Proceedings of a Symposium, Karlsruhe, 6-10 June 1966. Vienna: International Atomic Energy Agency.
6. WHO, 1981. Wholesomeness of Irradiated Food. Technical Report Series 659. Geneva: World Health Organization .
7. Diehl, J. F. and Josephson, E. S. 1994. "Assessment of wholesomeness of irradiated foods (a review)". Acta Alimentaria 23: 195-214.

8. Elias, P. S. and Cohen, A. J., eds. 1983. Recent Advances in Food Irradiation. Amsterdam: Elsevier Biomedical.
9. IAEA, 1989. Acceptance, Control of, and Trade in Irradiated Food. Conference Proceedings, Geneva, 12-16 December 1988. Vienna: International Atomic Energy Agency.
10. WHO, 1988. Food Irradiation. A Technique for Preserving and Improving the Safety of Food. Geneva: World Health Organization.
11. WHO, 1994. Safety and Nutritional Adequacy of Irradiated Food. Geneva: World Health Organization.
12. WHO, 1999. High Dose Irradiation of Food. Geneva: World Health Organization (in press).
13. Diehl, J. F., Hasselmann, C., and Kilcast, D. 1991. "Regulation of food irradiation in the European Community: is nutrition an issue?". Food Control 2: 212-219.
14. Satin, M. 1996. Food Irradiation: A Guidebook, 2nd. edit. Lancaster, PA: Technomic Publishing, Inc.

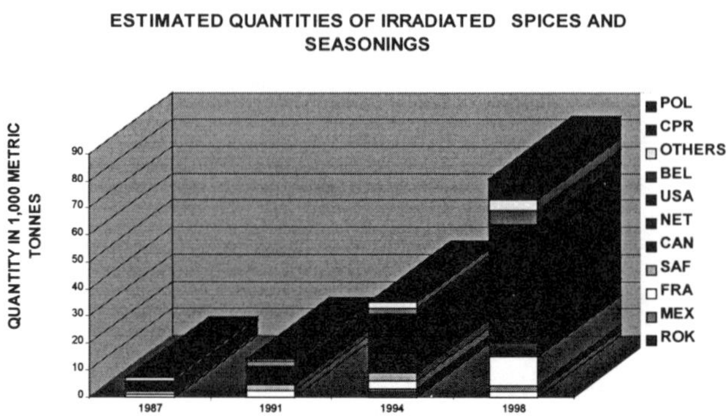

Figure 1: Estimated quantities of spices and dried seasonings irradiated in various contries. (source: IAEA)

Challenges and Opportunities for Food Irradiation in the 21st Century

L. M. CRAWFORD
Director, Center for Food and Nutrition Policy
Georgetown University
Washington, DC, USA

ABSTRACT

Writing in the Journal of the American Medical Association in 1994, Dr. Philip R. Lee, US Assistant Secretary of Health, asserted, "Food irradiation, like pasteurization of milk, can prevent countless infections because it destroys the pathogens that cause foodborne illness. The technology of food irradiation has languished too long already." A leading expert on the subject of food irradiation, Dennis Olson, concluded last year, "Irradiation of food can effectively reduce or eliminate pathogens and spoilage microorganisms while maintaining wholesomeness and sensory quality." This paper will address opportunities for food irradiation in controlling foodborne disease and preserving quality in the world to come. It likewise will address the challenge represented by consumer acceptance and regulatory approval.

Foodborne disease is a leading cause of morbidity and mortality around the globe. Hunger likewise is a constant companion to a significant number of the World's peoples. Food irradiation can significantly ameliorate both these problems. The destruction of pathogenic microorganisms in food by gamma irradiation is obvious but providing more food is less obvious. Approximately one-third of the world's food supply is either thrown out or allowed to spoil each day. Food irradiation can extend the shelf life of food by as much as twofold and thus would directly increase the availability.

The principal challenge of food irradiation for the future is to achieve mass production status in highly visible situations. This can be done by large volume usage in Western countries or in selected developing countries. The need for this relates to consumer acceptance. As long as utilization is minimal and in circumscribed areas, food irradiation will continue to be viewed as a novelty and/or an experiment when in fact it has been around for over 90 years.

Competition also can and would facilitate acceptance of the technology. In the US for many years, only one company was in the food irradiation business. This stymies innovation and entrepreneurship and continues to foster the public opinion that food irradiation is a specialty technique that is not applicable to a significant portion of the food supply. Finally a period of intense usage of food irradiation will doubtlessly result in fewer cases of food borne disease. These results will encourage wider employment of food irradiation and will serve to make the technology appealing to other nations.

The future of food irradiation is filled with promise. Consumers of the next century likely will grow to appreciate the technology for the lifesaving modality that it is.

Introduction

Writing in the Journal of the American Medical Association in 1994, Dr. Philip R. Lee, US Assistant Secretary of Health, asserted, "Food irradiation, like pasteurization of milk, can prevent countless infections because it destroys the pathogens that cause foodborne illness. The technology of food irradiation has languished too long already." A leading expert on the subject of food irradiation, Dennis Olson, concluded last year, "Irradiation of food can effectively reduce or eliminate pathogens and spoilage microorganisms while maintaining wholesomeness and sensory quality." This paper will address opportunities for food irradiation in controlling foodborne disease and preserving quality in the world to come. It likewise will address the challenge represented by consumer acceptance and regulatory approval.

Foodborne disease is a leading cause of morbidity and mortality around the globe. Hunger likewise is a constant companion to a significant number of the world's peoples. Food irradiation can significantly ameliorate both these problems. The destruction of pathogenic microorganisms in food by gamma irradiation is obvious but providing more food is less obvious. Approximately one-third of the world's food supply is either thrown out or allowed to spoil each day. Food irradiation can extend the shelf life of food by as much as twofold and thus would directly increase the availability.

The principal challenge of food irradiation for the future is to achieve mass production status in highly visible situations. This can be done by large volume usage in Western countries or in selected developing countries. The need for this relates to consumer acceptance. As long as utilization is minimal and in circumscribed areas, food irradiation will continue to be viewed as a novelty and/or an experiment when in fact it has been around for over 90 years.

Competition also can and would facilitate acceptance of the technology. In the US for many years, only one company was in the food irradiation business. This stymies innovation and entrepreneurship and continues to foster the public opinion that food irradiation is a specialty technique that is not applicable to a significant portion of the food supply.

Finally a period of intense usage of food irradiation will doubtlessly result in fewer cases of food borne disease. These results will encourage wider employment of food irradiation and will serve to make the technology appealing to other nations.

The future of food irradiation is filled with promise. Consumers of the next century likely will grow to appreciate the technology for the lifesaving modality that it is.

Breakthrough in the US

In late September Tyson Foods, Inc. and the Titan Corporation announced an agreement on food irradiation that is of great historical significance. Titan will provide its patented Surebeam technology to Tyson on an exclusive basis. This includes electron beam, x-ray or any other type of ionizing energy. Test marketing will commence in the Spring of 2000.

Titan, a San Diego corporation, is building an irradiation facility in Sioux City, Iowa. They already have commitments from IBP and Cargill Foods. The Titan technology is approved by FDA and FSIS.

Electron beam acceleration is widely used in France, particularly on mechanically separated poultry products. This technology involves the generation of electrons from a commercial electrical source; the beam of electrons is then targeted, usually through a system of revolving heads, onto the food for a prescribed length of time. X-rays are generated in much the same way but require a great deal more power and are therefore much more expensive. The third modality, gamma irradiation, polarizes public opinion and thus was specifically excluded from consideration by a Tyson spokesperson.

The official statement from Tyson Foods is a document remarkable in its candor and forthrightness. It clearly shows a preference for the term cold pasteurization although the first use of the term is followed by irradiation in parentheses. The statement also points out that the process both dramatically reduces harmful bacteria and can add valuable days to shelf life. It then frankly states that this is not the single source solution to safe food.

Tyson statement is most impressive in its discussion of endorsing clear communication to the consumer through information panels rather than warning labels concluding that eventually labelling should be made optional.

Finally, Tyson reports on the medical and scientific endorsement of the World Health Organization, the American Medical Association and the American Dietetic Association. And it registers the corporations firm belief that long-term consumer demand and the markets ability to support the added cost of this technology will be the key to future expansion of the entire food industry. This just after revealing that they only intend to offer irradiated products as alternatives to some of their existing products.

Tyson is the nations largest poultry producer and is well respected for its leadership position in the food industry. This unqualified and informative endorsement of food irradiation may be the break-through that food safety experts have long prayed for. Even limited use by Tyson will be good for public health and a no nonsense statement to those who have unfairly belittled this lifesaving technology.

What a stark contrast the aforementioned epochal events are to the so-called organic movement. At long last, a technology that will improve the lives and well being of virtually everyone as opposed to a fad that, at the end of the day, offers only a promise of returning to the Garden of Eden.

Role of WHO

WHO continues to provide international leadership, guidance and education regarding food irradiation. The 1999 report, High-Dose Irradiation: Wholesomeness of Food Irradiated with Doses Above 10 kGy, was particularly helpful. The following excerpt is illustrative, food irradiated to any dose appropriate to achieve the intended technological objective is both safe to consume and nutritionally adequate. And irradiation to high doses is essentially analogous to conventional thermal processing, such as the canning of low-acid foods, in that it eliminates biological hazards from food materials intended for human consumption, but does not result in the formation of physical or chemical entities that could constitute a hazard.

Organizations as prestigious as the American Medical Association have been able to endorse food irradiation based on the imprimatur of WHO. The sustained and informed endorsement of food irradiation by WHO have set the stage for the global acceptance of food irradiation. The remainder of this presentation will dwell on the means whereby that acceptance can be accomplished.

Opportunities

Perhaps no food process has been researched and reviewed more than food irradiation. The scientific basis and mechanisms of food irradiation are as thoroughly understood by the scientific community and regulatory agencies as traditional food processes. While irradiation provides, it cannot replace proper food handling as the single most critical food safety measure. Irradiation does not prevent contamination but it controls it.

A broad spectrum of national and international studies conclude that food irradiation can accomplish the following:

Inhibit sprouting in crops like potatoes, onions and garlic;
Destroy insects and parasites in cereal grains, dried beans, dried and fresh fruits, meat and seafood;
Delay the ripening and spoilage of fresh fruits and vegetables;
Extend the shelf-life of perishable products like beef, poultry and seafood;
Eliminate disease-causing microorganisms in food; bacterial and fungal spores and viruses are not affected by the most commonly used dose levels; and
Sterilize at doses above 10 kGy herbs, spices and other ingredients like dried vegetables, in addition to foods prescribed for immuno-compromised hospital patients and for astronauts during space flight.

There is a large body of scientific evidence concerning the effects of irradiation on microorganisms found in food. The most common and most pathogenic of these bacteria *Salmonella*, *Shigella, Yersinia, Campylobacter* And *Listeria* are most sensitive to radiation and can be readily eliminated by the process at the usually employed doses. Medium dose irradiation is also effective in the control of *Escherichia coli* 0157:H7. This is likely true for the other toxigenic strains of *E. coli.* Irradiation also has proved effective against *Staphylococcus aureus,* which is another source of food contamination in meat.

Importantly, microflora surviving low or medium dose irradiation generally include spoilage bacteria which are useful in gauging wholesomeness. Food irradiation is recognized by the world health community as another method for preserving food and ensuring its wholesomeness. In the last 50 years, research activities have examined this process and concluded that when used under good manufacturing practices, irradiation will increase food safety and reduce foodborne illness, lengthen shelf-life of many perishable products, and reduce the risks of post-processing contamination of intact, packaged products.

The incipient opportunity inherent in food irradiation is that of a safer food supply and a world that is much more free of foodborne disease. It would be irresponsible not to pursue that opportunity fully with all reasonable diligence.

Challenges

There are a number of formidable challenges to the full utilization of food irradiation. These are:
Consumer acceptance;
Growth of the irradiation industry;
Logistics of the source;
Erection of a Research and Development Program;
Regulatory Reform.

<u>Consumer Acceptance</u>

Opinion polls vary somewhat but generally indicate approximately 25-30% of consumers have initially favorable impressions of food irradiation, 55-65% are uncertain and 5-10% are opposed. Some polls taken in the 1990s show that a majority of consumers would favor such items as irradiated meat and would be willing to pay a premium for the product.

Actual market tests have found that consumers are inclined to purchase irradiated foods after they have been educated about the safety and benefits of the irradiation process. In-store promotion of various items including chicken, strawberries, mangoes, papayas and other items in separate tests have proved this often with dramatic sales results in favor of the irradiated product.

Surveys generally indicate that when consumers are informed about the irradiation process and its uses in food, the acceptance of both the theory and practice of irradiating foods rises markedly. Earlier studies that reported higher numbers of people expressing major concern about irradiation were often surveys that did not also provide information about the process. The US Food Marketing Institute has found that 36% of people surveyed were very or somewhat likely to purchase irradiated foods after being given a brief description of potential benefits

Irrespective of all these studies and despite the intuitive good inherent in food irradiation, the fact remains that consumers are not demanding that food be irradiated. The situation therefore is that food purchasers may acquiesce to the utility of food irradiation but there is no active consumer-driven campaign in support of irradiated food. Therefore, the current situation is not likely to result in widespread consumption of irradiated products.

The key to changing this situation is to join sales promotion and product launch campaigns with effective education programs. Within the past few weeks, Dr. Elizabeth Whelan of the

American Council for Science and Health in New York City, has broadcast a radio message in most of the major markets in the US extolling food irradiation. This appears to have been effective, but this kind of initiative clearly needs to be sustained and it needs to be individually targeted for individual markets. The use of celebrities including well-known medical and scientific figures in addition to Dr. Whelan would likewise be advantageous.

The central message of the aforementioned campaign should be that food irradiation is a safe modality that will save lives, reduce illness and disablement as well as extending shelf-life. The message needs to be continually re-worked and re-packaged and should regularly convey different aspects of the safe food through irradiation core presentation. And, most of all, every iteration should ring with candor, forthrightness and the conveyance of new information branching out into other areas of food safety and even nutrition.

Growth of the Irradiation Industry

At the present time, most countries have either one or only a few irradiation facilities. Worse still, in the majority of countries that irradiate to any extent there is only one or two companies or government entities involved in irradiation. This fosters the impression that food irradiation is a fledge line, untried enterprise that may not be viable. And it breeds the consumer wait-and-see attitude.

What is needed is a robust, competitive industry that vigorously promotes and develops the technology. Trade associations representing the food irradiation industry could do a great deal of good for consumer acceptance and government support.

A rapidly expanding industry would quite naturally move toward research and development aimed toward improved delivery systems, better packaging and more effective public relations. Countries should consider ways and means of promoting growth in the food irradiation industry through tax advantages, subsidies, development grants and deregulation.

Logistics of the Source

The current tendency toward a centrally located facility where food is taken to be irradiated is all wrong and will impede further progress until it is abandoned. The irradiation source must be either in or adjacent to the food plant not in some other location where the food must be transported.

The appeal of electron beam acceleration is in no small part due to the potential for applying the technology on-line. This would offer the advantage of not having to transport the food to and from a central facility with the attendant potential for re-contamination and reduced shelf-life.

Central, dedicated facilities are inconvenient, awkward and they unnecessarily bring attention to irradiation as an entity separable from the safe food it creates. Central facilities also add extraordinary expense because of remoteness and the need for a separate building and the infrastructure demanded by that building.

In short, the concept of a central facility is incompatible with further expansion of the food irradiation industry. Different, more radical approaches need to and must be developed in lieu of the central facility.

Erection of a Research and Development Program

What is the need for a research and development program if the technology is safe and efficacious? Because continuous improvement and innovation must mark the future of food irradiation.

Presently, some foods with high fat contents or other physico-chemical properties are not good candidates for irradiation because of undesirable changes in sensory characteristics. These foods should be researched and alternative approaches with varying levels or varying exposure times might well result.

Moreover, now that it is generally accepted that high doses of irradiation can be employed with impunity, varying levels and exposure-times should be the subject of experimentation with all foods. Exquisite calibration of the dose and duration will enable a better understanding of the precise circumstances under which certain pathogens can be reduced or eliminated. Also, the control of parasites and perhaps some viruses should be subject of further experimentation.

Finally, well-designed studies aimed at precisely describing the optimum levels for the most desirable extension of shelf-life should likewise be the subject of experimentation. Secondary sensory characteristics such as coloration, smell, taste and appearance should also be researched.

Unique usage of irradiation such as the sterilization of beverage waters, food supplements and food packaging are in need of carefully designed studies with endpoint characterization. The utility and manifold indications for irradiation are virtually endless. Much research to this point has been devoted to establishing safety. It is now time to move past this in an exploration of newer and more precise uses of the technology.

Regulatory Reform

In many countries, there has been erected a cumbersome wall of regulations, punitive labelling and unnecessary warnings that discourages the use of food irradiation. The use of the radura symbol is particularly opprobrious in that it resembles a mushroom cloud. And to require verbiage in addition to the radura symbol is clearly counterproductive. Furthermore, many countries require that each and every foodstuff to be irradiated be the subject of approval processes designed as if the procedure had not previously been the subject of an approval.

In the US, irradiation was approved for pork in 1986, for chicken in 1992 and may be approved for beef in 1999. The approval process required 6 years for pork, 6 years for chicken and 7 years or longer for beef. This is patently absurd. Food irradiation should either be outlawed or approved. The ridiculous delays in the approval process is antithetical to public health.

A blanket approval for food irradiation should be sought in every country. This term implies that the process be approved with appropriate safeguards for the partial sterilization and shelf-life extension of food. After this, companies and others seeking governmental permission to employ the process should be required to register with the government listing all foods that it intends to irradiate, and be subject to an on-site inspection by appropriate officials.

All containers of irradiated food should be identified with a written legend such as this food has been irradiated. The radura symbol should be eliminated. Irradiated food should not be segregated except as is necessary to prevent re-contamination.

Government regulatory bodies should be stimulated to develop policies that encourage the use of irradiation. And such bodies should disseminate helpful information that demonstrates the effectiveness and utility of food irradiation.

Conclusion

All attendees at this Conference should go forth from here with a renewed commitment to be proponents of food irradiation and to everywhere stimulate its use. The various steps that must be taken in each country to finally bring about the acceptance of food irradiation should be the partial responsibility of its scientific and medical establishment. Thus it is our duty to our disciplines and to our countries to be available to constructively support the demystification of food irradiation and the inculcation of the technology into the culture of all the nations of the world.

Harmonisation of Regulations on Food Irradiation in Accordance with the SPS and TBT Agreements

P. B. ROBERTS
Institute of Geological & Nuclear Sciences
Lower Hutt, New Zealand

ABSTRACT

The approach of the new millennium brings with it an increasing prospect for international trade in irradiated foods. This prospect should re-focus attention on the need for regulations on irradiated foods to be harmonized as much as possible. Without harmonization, international trade may be significantly hindered.

Through its ability to produce safe, hygienic foods, irradiation can assist developing countries meet the strict requirements for entry into markets in developed countries. Trade in herbs and spices and in fruit and vegetables are examples of commodities for which irradiation can guarantee safety from microbial contamination and insect pests. Irradiation is now a favoured treatment option for such foods as it eliminates the need for treatments by chemicals that are of health and environmental concern. Other options for trade in irradiated foods include meat and seafoods. Notable is recent action which will lead to the commercialisation of the irradiation of red meat in the United States and increasing concerns generally about contaminants such as *E. coli* O157:H7 and *Listeria monocytogenes*.

Over forty countries have regulations in place permitting the irradiation of one or, frequently, many foods. However, overall, there is relatively little uniformity. In the regulatory environment for irradiated foods. This is despite the existence of a Codex General Standard for Irradiated Foods since 1984 and the responsibility placed on member states of the World Trade Organisation to harmonize sanitary and phytosanitary measures on as wide a base as possible.

The International Consultative Group on Food Irradiation has published Guidelines for the Authorisation of Food Irradiation Generally or by Classes of Food. This recommends a framework of harmonized regulations based on the Codex General Standard. In the area of phytosanitary measures, regional plant protection authorities have endorsed irradiation as a quarantine treatment of fresh fruits and vegetables. Steps to provide a harmonized framework for irradiation as a quarantine treatment have recently been made in countries in Asia and the Pacific. The progress made towards harmonized regulations in the regions of Asia/Pacific, Latin America, Africa and the Middle East will be reviewed.

INTRODUCTION

There are few internationally recognised standards and agreements related directly to irradiated foods. The most important of these have been produced by the Codex Alimentarius Commission. Codex is the recognised international authority for standards related to food safety and human health. This article describes the basis and implications for national regulations of the Codex General Standard for Irradiated Foods. Progress towards harmonisation of national regulations and regulations within regions will be discussed.

International trade in irradiated foods is still of very low volume at present. The World Trade Organisation has put in place two major Agreements to govern international trade. Although neither the Sanitary and Phytosanitary Agreement nor the Technical Barriers to Trade Agreement directly refer to irradiated foods, both Agreements may have significant implications for attitudes to regulations that may govern future international trade in irradiated foods.

CODEX STANDARDS AND IRRADIATED FOODS

The large database on the safety and nutritional adequacy of irradiated foods has been evaluated by many national and international committees. The findings of these committees can be summarised by those of a 1980 FAO/WHO/IAEA Joint Expert Committee on Food Irradiation (1). The findings were that irradiation of any commodity up to an overall average absorbed dose of 10 kGy presents no toxicological hazard. It also introduces no special nutritional or microbiological problems. These findings were confirmed once again in 1994 in a special WHO report.

Based on the findings, in 1984 Codex issued a World-Wide General Standard for Irradiated Food (2). It also issued a Recommended Code of Practice for the Operation of Radiation Facilities for the Treatment of Food (3). The Codex General Standard sets conditions that, in brief, include :

(a) the radiation sources that may be used. These are cobalt-60, cesium-137, x-rays below 5MeV and electrons below 10MeV;
(b) absorbed dose. The overall average absorbed dose shall be less than 10 kGy;
(c) hygiene. The process must comply with Codex principles and should not be a substitute for Good Manufacturing Practice;
(d) labelling. There should be standard shipping documentation and labelling should comply with Codex principles.

On labelling, the Codex General Standard for Labelling of Pre-packaged Foods was revised in 1991 (4) and has three clauses on irradiated foods -:

5.2.1 The label of a food which has been treated with ionizing radiation shall carry a written statement indicating that treatment in close proximity to the name of the food. The use of the international food irradiation symbol is optional but when it is used, it shall be in close proximity to the name of the food.

5.2.2 When an irradiated product is used as an ingredient in another food, this shall be so declared in the list of ingredients.

5.2.3 When a single ingredient product is prepared from a raw material which has been irradiated, the label of the product shall contain a statement indicating the treatment.

Earlier this year, the Codex Committee on Food Additives and Contaminants accepted the request of the International Consultative Group on Food Irradiation (ICGFI) and initiated a process of revision of the Codex General Standard, particularly with respect to the maximum absorbed dose stipulation. This action followed the findings of a WHO Study Group that met in September 1997 to consider the safety of foods irradiated with doses above 10 kGy (5). Their conclusion was that foods treated with doses above 10 kGy can be considered safe and nutritionally adequate when produced under conditions of Good Manufacturing Practice. The Study Group believes that there is no need for any upper limit on the dose applied to a food from the point of view of safety. The upper limit can be set by the technological requirements to provide a product that satisfies the consumer's requirements of taste, texture, smell and appearance.

However, the Codex Committee on Labelling decided at its Session in Ottawa in April 1999 to take no action on the General Standard on Labelling on Pre-packaged Foods as it relates to irradiated foods and especially irradiated ingredients.

NATIONAL REGULATIONS AND HARMONISATION

Present Situation

A major feature of the Codex General Standard is that it treats irradiation as a physical process. It does not refer to any specific food or use of the process. The implication is that irradiation should be regulated in the same way as other physical treatments of food such as canning, pasteurisation and microwaving.

Most countries require the irradiation facility to be licensed for radiation safety by a specialised government agency. Responsibility for regulating the application to foodstuffs is then usually in the hands of a different agency, such as a Ministry of Food or Public Health.

Forty-one countries have informed the Joint Division of the FAO/IAEA that they have listed at least one type of food as cleared for irradiation for one or more purposes (6). Although the Codex General Standard declares that <u>any</u> food can be irradiated up to an overall average dose of 10 kGy, most countries regulate irradiation using the principles of food additives. They still require each food or application to be approved on a case-by-case basis. This philosophy is a basic cause of non-harmonised regulations.

The listing of countries and approved foods appears extensive. However, the link between these lists and actual regulations, final authorisations and true commercial applications is not simple.

Some countries have specific regulations on food irradiation, and the list of approved items is part of the regulation. In other countries the list of approved items has been drawn up by a statutory agency or special committee. There is no specific food irradiation regulation but an authorisation can be given within general food regulations.

It is also important to note that the existence of an approval does not necessarily mean that there is automatic authorisation to proceed with commercial processing. Usually it is still necessary to make a specific application to process food, to provide justification and to show that good manufacturing practice will be carried out.

Two aspects of national regulations, in particular, may cause a lack of uniformity. One is labelling, especially on the labelling of food ingredients. The Codex General Standard on Labelling is being interpreted in different ways. Some countries require no labelling of irradiated foodstuffs once the food becomes an ingredient of another food; others require labelling regardless of how small a proportion of the complete food may be comprised of an irradiated ingredient; other countries make no reference to labelling.

The other cause of non-uniformity concerns the allowable doses of radiation applied to the food. The Codex Standard allows irradiation of food up to a maximum of 10 kGy overall overage dose. This maximum is based on very conservative safety and nutritional considerations. In practice, it is also necessary for processors to limit the dose range applied to foods for technological reasons. These reasons include ensuring that the intended benefit is provided, for which a minimum dose is necessary, and retaining the quality of the food, for which a maximum below 10 kGy may be necessary. These technological limits, which are basically advisory and the business of the radiation processor, are often included in national regulations in different ways in different countries.

Progress Towards Greater Harmonisation

In 1994 ICGFI published Guidelines for the Authorisation of Food Irradiation Generally or by Classes of Food (7). This recommends and describes a framework for harmonised regulations based on the Codex General Standard. A specific suggestion is to regulate foods by major food classes rather than individual foods as an intermediate step towards the Codex position.

In 1993 a Model Regulation for Asia and the Pacific was developed by food control officials. Officials from the ASEAN countries used the Model Regulation to produce a draft Harmonized Regulation on Food Irradiation that was considered by ASEAN Ministers of Agriculture and Forestries.

In April 1998 a meeting attended by senior officials representing food safety, agriculture and nuclear technology in 14 Asia / Pacific countries took place in Seoul, Korea. The Seoul meeting considered the ASEAN draft and revised it to produce a Harmonised Regulation on Food Irradiation that is available from ICGFI (7).

The result was a recommendation to base national regulations on the Harmonised Regulation as soon as feasible. At the end of 1998, this recommendation was forwarded for consideration by all 17 Member States within the Regional Co-operative Agreement (RCA) on Nuclear Science.

There has already been some action on the initiative within the Asia/Pacific region. ASEAN Ministers of Agriculture and Forestries have issued an instruction to regulate irradiated foods according to the Harmonised Regulation, that is, as a food process according to the Codex General Standard for Irradiated Foods. Bangladesh, Pakistan and China now deal with foods on the basis of a few food classes rather than on an item by item basis. For many countries in the region, the over-riding determinant of approval is whether the procedure meets the criteria of the Codex General Standard.

In Singapore trade in irradiated food is controlled under Chapter 238, Article 38 of the Sale of Food Act, "Food Regulations". Under the regulation, food irradiation is only permitted under licence. But in granting a licence there is no specific consideration of the food / dose / purpose. Rather it is required that :

the irradiation is conducted in accordance with the Codex Code of Practice;
irradiated food meets the Codex General Standard for Irradiated Foods;
Codex labelling requirements are followed.

There is definite movement towards an easier and more consistent regulatory environment for food irradiation in Asia. This movement is being continued in other regions. A similar Model Regulation has been recommended for use in Latin America, in Africa and in the Near East, all based around the ICGFI Guidelines and Codex General Standard. The Harmonised Regulation for Asia/Pacific and the Model Regulation for the Near East Countries are noteworthy in that they no longer follow Codex in one respect. They recognise the work of the WHO Study Group on High Dose Irradiation and eliminate all reference to a limit of an overall average absorbed dose of 10 kGy. They introduce instead "Advisory Technological Dose Limits" to guide but not limit irradiation processing of food.

Countries outside Asia that have indicated that they have or may soon move to a regulatory environment in conformity with the Codex General Standard are Argentina, Brazil, Chile, Morocco, Peru, Ghana and South Africa. Mexico has for some time regulated irradiated foods according to food classes.

Among the developed countries, the United Kingdom has a food irradiation regulation since 1991 that allows applications to irradiate foods to be made within 7 broad food classes rather than by individual specific foods. The United States is also moving in this direction. However, an EU Directive allows only spice irradiation in EU countries starting September 2000.

The Future for Harmonisation

Irradiation has the potential to increase trade between countries through its ability to ensure that foods meet increasingly strict hygiene and bio-security (quarantine) standards and to promote a future less dependent upon chemical treatments that are known health and environmental risks.

However, if this potential is to be met, then the barriers presented by non-harmonised national regulations must be addressed. As I have already described, there is progress towards greater harmonisation based on the ICGFI Guidelines, Model Regulations for regions and the Asia/Pacific Harmonised Regulation.

At this stage in the commercial development of food irradiation, some governments may still choose a quite cautious approach and require consideration of irradiated foods on a case-by-case basis. The result may be a regulation that is a basic ban on the process but which permits applications to be made for specific exemptions. In this situation, the guidelines governing how an exemption can be granted should be in conformity with the Codex General Standard.

WORLD TRADE ORGANISATION AGREEMENTS

At present the volume of irradiated food moving in international trade is very small. If this situation is to alter in the New Millennium, then the Agreements administered by the World Trade Organisation should be considered.

The WTO was established as a successor to the General Agreement on Tariffs and Trade (GATT). The final round of talks under GATT, the Uruguay round, paid far greater attention to food and agricultural commodities than previous rounds. The Uruguay round was completed in December 1993 and involved several individual agreements. The purpose of the talks was to liberalise global trade and, particularly, to minimise the use of technical (non-tariff) barriers to trade.

Governments that ratify the Agreement to establish the WTO automatically agree to be bound by all multi-lateral trade agreements attached to the main Agreement. The Agreement sets out procedures to settle trade disputes. However, only when a trade barrier has been challenged and ruled upon can the exact meaning or legal standing of clauses within Agreements be resolved. As no formal disputes arising from trade in irradiated foods have occurred, any interpretation of the Agreements has to be tentative at this stage.

General Principles

The WTO Agreements -

- maintain the sovereign rights of Governments to take all necessary measures to meet legitimate objectives (e.g. national security, health, prevention of deceptive practices).
- forbid misuse of measures for protectionism or for creating unnecessary trade barriers.
- affirm that levels of protection (e.g. for health) cannot be challenged; however, measures taken to achieve that level can be challenged.
- demand that measures shall not be more trade restrictive than necessary to achieve the legitimate objective; the least restrictive measure should always be taken.

Trade in food and agricultural commodities may be affected by both the principal Agreements within the overall WTO Agreement, neither of which specifically refer to irradiation or irradiated foods. The principal Agreements are the -:

(1) Technical Barriers to Trade (TBT) Agreement

This concerns any measure relating to a wide range of legitimate objectives. Examples would be regulations on labelling, deceptive practices, consumer rights. Issues such as regulations concerning extension of shelf-life, storage, organoleptic properties of food would come under TBT.

(2) Sanitary and Phytosanitary (SPS) Agreement

This concerns any measure related to the specific objective of protecting the health of human, animal or plant life. The SPS could be regarded as a special exception to TBT. As a more specific Agreement, the SPS takes precedence over the TBT.

SPS Agreement

The SPS Agreement recognises the relevant international standards, guidelines and recommendations of competent organisations, including the :

Codex Alimentarius Commission (for Food Safety and Human Health);
International Office of Epizootics (for Animal Health);
International Plant Protection Convention (for Plant Health).

We have already noted that Codex has a World-Wide General Standard for Irradiated Food. Also regional organisations within IPPC, such as the North American Plant Protection Organisation, have recognised irradiation as a technically viable and effective option for insect disinfestation for quarantine purposes.

The SPS Agreement has as its basis :

(a) the right of Governments to establish the level of protection they consider necessary;
(b) national measures shall be based on international standards, guidelines and recommendations where these exist, and on scientific evidence and risk assessment;
(c) countries can take measures more stringent than international standards etc, if they have scientific evidence, or if the standards do not provide their acceptable level of protection, as shown by rigorous risk analysis procedures.
(d) the equivalence of different measures to achieve the level of protection is to be recognised;
(e) decisions on acceptable level of risk cannot be varied from situation to situation (or measure to measure);
(f) importers may take precautionary measures while awaiting further scientific evidence;
(g) exporters can request reasons, or ultimately challenge, if restrictive measures are not based on international standards.

There is a general consensus that trade restrictions on irradiated foods based on a need to defend public health could be difficult to justify. This is because of the requirement of the SPS Agreement to account for international standards and scientific evidence. There is a vast amount of scientific data on the safety of irradiated food and its efficacy as a plant quarantine treatment plus a Codex General Standard that permits irradiation of any food up to an overall average absorbed dose of 10 kGy.

TBT Agreement

The TBT Agreement requires :

(a) relevant scientific data and risk assessment to be taken into account;
(b) deviations from relevant international standards to be explained if requested;
(c) any measure taken to be the least restrictive to trade possible; note that a ban is usually the most restrictive measure possible.

There is a general consensus that as scientific evidence is only to be considered and is not a mandatory requirement, restrictions falling within the TBT will be easier to defend than SPS-related restrictions. However, it should be noted that consumer issues generally can be best dealt with by labelling requirements rather than by bans, and that there are international labelling standards.

CONCLUSIONS

The SPS and TBT Agreements are part of international efforts to liberalise world trade. Measures restricting irradiated foods that -

- have no scientific justification;
- are more stringent than for competing processes when subject to a risk analysis;
- could be replaced by less trade restrictive measures;

have the potential to be challenged under the SPS and TBT Agreements.

Trade in irradiated foods will also be encouraged by national regulations that are more uniform and harmonised. There are signs that many countries are at last moving towards a more harmonised system based around the ICGFI Guidelines for the Authorisation of Food Irradiation Generally or by Classes of Food and, thereby, more in conformity with the Codex General Standard.

When a national and international framework is in place for the easy sale/export/import of irradiated foods, then we shall finally obtain a truer picture of the real demand for the process in the market place. We shall discover at last whether, as some critics maintain, irradiation is an unnecessary and unwanted process. Or whether, as the technical and scientific data show, it will take its place alongside pasteurisation of liquid foods as an important tool in the production of safe, nutritious food.

REFERENCES

[1] WHO 1981. Wholesomeness of irradiated food. Report of a Joint FAO/IAEA/WHO Expert Committee. WHO Technical Report Series, No. 659. World Health Organisation, Geneva.

[2] Codex 1984a. Codex General Standard for Irradiated Foods (Codex STAN 106-1983). Codex Alimentarius Commission, Vol. XV, Geneva.

[3] Codex 1984b. Codex Recommended International Code of Practice for the Operation of Radiation Facilities used for the Treatment of Foods (CAC/RCP 10-1979, Rev. 1). Codex Alimentarius Commission, Vol. XV, Geneva.

[4] Codex (1985) Codex General Standard for the Labelling of Pre-packaged Foods (Codex STAN 1-1985, Rev. 1-1991). Codex Alimentarius Commission. Food and Agriculture Organisation of the United Nations, Rome.

[5] WHO 1994. Wholesomeness of foods irradiated above 10 kGy. Report of a WHO Study Group. World Health Organisation, Geneva.

[6] Anon 1998. "Clearance of item by country," Food and Environmental Protection Newsletter, Vol. 1, No. 2, IAEA, Vienna, Supplement, pp. 1-17.

[7] ICGFI 1994. Guidelines for the Authorisation of Food Irradiation Generally or by Classes of Food. ICGFI Document 15. International Consultative Group on Food Irradiation, IAEA, Vienna.

[8] ICGFI 1998. "Harmonisation of Procedures and Regulation on Food Irradiation for Asia and the Pacific," Proceedings of a Workshop, Seoul, Korea, 27-29 April, 1998. International Consultative Group on Food Irradiation, IAEA, Vienna.

Food Irradiation Harmonization in the European Union

H. VOUNAKIS
Principal Administrator, Foodstuffs Unit (E-1)
European Commission

ABSTRACT

On 22 February 1999, the European Parliament (EP) and the Council adopted formally two legislative texts regulating the food irradiation in the European Union. The agreed texts, published in the EU Official Journal L n' 66 of 13 March, consist of:-

a. **A framework** Directive on the general and technical aspects for carrying out the process, on labelling and decisions to authorise foods and food ingredients treated with ionising radiation, and

b. **An implementing** Directive establishing a **Community list** of foodstuffs authorised for irradiation treatment.

At the adoption, the list contained a single food category, *the dried aromatic herbs, spices and vegetable seasonings.* EU countries have to permit the marketing and use of these products by September of 2000.

The objectives of the framework directive are:

-- to lay down the rules to be observed by the units carrying out irradiation (standards, inspections, records, radiation sources, etc.);

-- to define the conditions for authorising foodstuff irradiation (purposes, wholesomeness of food, reasonable need for treatment);

-- to establish procedures for authorisation of treatment with ionising radiation and the marketing and import of foods treated (doses and sources of radiation, inspections, labelling, etc.

Their legal basis is Article I OOA of the Treaty, which foresees that amendments related to the above issues are to be carried out by a joint Council and Eur. Parliament decision procedure, (codecision), which usually requires at least one and a half years.

Both directives also provide a *safeguard clause,* which allows the Eur. Commission to use a shorter procedure in order to adopt quickly amendments that become necessary following the introduction of urgent national measures. This is a widely used standard clause, aiming only at the protection of public health under urgent circumstances.

The points of the framework directive presenting a particular interest for trade are:

- the establishment of the Community list of irradiated products in stages (Art. 4.2),
- the conditions under which existing national authorisations would operate in transitional period (Article 4.4).
- the compulsory labelling of any irradiated food and food ingredient, even when incorporated in compound ingredients used in finished products (Art. 6. 1 c),
- the official approval of irradiation installations of any country (Art. 7 and 9)

For the completion of the above list the Eur. Commission shall make a proposal to the EP and the Council by the end of year 2000 (Art. 4.3), taking into account existing national authorisations and the opinion of the Scientific Committee for Food (SCF). A letter has been sent to this purpose to Member States, asking for suggestions. The decision on the eventual transfer of national authorisations to the EU list will be taken by **codecision.**

National authorisations existing at the time of adoption may be maintained until the entry into force of the completed list foreseen by Art. 4.3, provided they are cleared by the SCF and comply with the framework directive; otherwise they cease to be valid (Art 4.4). No national authorisation can be issued any more. The restriction of issuing new authorisations does not apply for products that had been authorised in other Member States and cleared by the SCF (Para. 5).

Until the entry into force of the directive completing the Community list, Member States may retain their **national bans,** *in compliance with the general rules of the EU Treaty* (Para. 7).

Irradiated foodstuffs imported from third countries are subject to the same conditions regulating their treatment, as in the European Union. Their imports may be permitted if the irradiation facilities carrying out the treatment in the country of origin have been approved by EU. The list of irradiation facilities which would be approved, shall be published in the *Official Journal of the European Communities.*

Preparatory work on the approval procedure is expected to start by next November.

On 22 February 1999, the European Parliament (EP) and the Council adopted formally two legislative texts regulating the food irradiation in the European Union. Their legal basis was the ex Article 100A of the Treaty aiming at the completion of the internal market, renumbered to Art. 95 after the signature of the Amsterdam Treaty. The agreed texts, published in the EU Official Journal L n° 66 of 13 March, consist of:

a. A framework Directive on the general and technical aspects for carrying out the process, on labelling and decisions to authorise foods and food ingredients treated with ionising radiation, and
b. An implementing Directive establishing a *Community list* of foodstuffs authorised for irradiation treatment.

The need for harmonising EU legislation also in this area, had been acknowledged in the Commission's legislative programme of 1985 completing the internal market, the so called White Paper. It is worth stressing that the first working documents to this purpose had already been drafted by the Commission services in the 70's. However, opposing national legislations eliminated the chances for envisaging a proposal at that time. Only after a major change on food policy took place in two Member States was it possible to undertake discussions, even with a restrained optimism.

The Scientific Committee for Food (SCF), an independent body advising the Commission on health matters, was then asked to consider whether food irradiation might result in risks for humans. The Committee, assisted by four experts, examined more than five hundred bibliography references and endorsed *eight* food classes, setting technological limits for doses according to the table below. It delivered a favorable opinion (March 1986), which was published in the EUR series, in all Community languages. The SCF concluded that the process posed no problem for health over a wide range of uses, if used under prescribed conditions. It noted that an extension of the list could be envisaged, provided that appropriate information were given for evaluation.

The table has been completed twice since then, first in 1992 to include *Camembert* cheeses manufactured from raw milk and again in 1998, when the Committee endorsed *eight* individual foodstuffs.

	Food classes	Overall Average Radiation Dose (kGy)
	Endorsed by the EU Scientific Committee for Food	
1.	Fruits	up to 2
2.	Vegetables	up to 1
3.	Cereals	up to 1
4.	Starchy tubers	up to 0.2
5.	Spices and condiments	up to 10
6.	Fish and shellfish	up to 3
7.	Fresh meats	up to 2
8.	Poultry	up to 7
9.	Camembert cheeses	up to 2,5
10.	Frog's legs	up to 5
11.	Shrimps	up to 5
12.	Gum arabic	up to 3
13.	Casein/caseinates	up to 6
14.	Egg White	up to 3
15.	Cereal flakes	up to 10
16.	Rice flour	up to 4
17.	Blood products	up to 10

The SCF report facilitated the submission to the Eur. Parliament and Council a Commission proposal for an EU directive (9.12.88), containing *eight food categories and three products.*

The debates with Member States in the Council on the Commission proposal made it clear that an agreement would not be possible unless the list of products to be authorised was reduced to the strict minimum of applications, for which treatment was necessary to protect public health. At the adoption of the directives , the list contained a single food category, *dried aromatic herbs, spices and vegetable seasonings.*

EU countries now have to permit the marketing and use of these products by September 2000. They also have to prohibit those foodstuffs not complying with the provisions of these Directives by 20 March 2001. The maximum overall average radiation dose that may be absorbed by products of that specific category has been fixed at 10 kGy.

OBJECTIVES OF THE EU LEGISLATION

The objectives of the framework directive are:

- to lay down the rules to be observed by irradiation units (standards, inspections, records, radiation sources, etc.);

- to define the conditions for authorising foodstuff irradiation (purposes, wholesomeness of food, reasonable need for treatment);

- to establish procedures for authorising the treatment with ionising radiation and the marketing and import of foods treated (doses and sources of radiation, inspections, labelling, etc.).

Amendments to the directives related to the above issues are to be carried out by a joint Council and European Parliament decision procedure, (codecision), which usually requires at least one and a half years.

Both directives also provide a *safeguard clause*, which allows the Eur. Commission to use a shorter procedure in order to adopt quickly amendments that become necessary following the introduction of urgent national measures. This is a widely used standard clause, aiming only at the protection of public health under urgent circumstances. Its potential use in this case should rather envisage a prohibiting action to already authorised irradiated foodstuffs.

The framework directive is composed of *seventeen* articles and *three* annexes.

Annex I fixes the conditions for authorising food irradiation and the purposes served.
Annex II defines the authorised sources of ionising radiation, and
Annex III describes the dosimetry technique and procedures for control purposes.

CONDITIONS FOR AUTHORISING FOOD IRRADIATION (*ANNEX I*)

1. Food irradiation may be authorised only if:

- there is a reasonable technological need,
- it presents no health hazard and is carried out under the conditions proposed,
- it is of benefit to the consumer,
- it is not used as a substitute for hygiene and health practices or for good manufacturing or agricultural practice.

2. Food irradiation may be used only for the following purposes:

- to reduce the incidence of food-borne disease by destroying pathogenic organisms,
- to reduce spoilage of foodstuffs by retarding or arresting decay processes and destroying spoilage organisms,
- to reduce loss of foodstuffs by premature ripening, germination or sprouting,
- to rid foodstuffs of organisms harmful to plant or plant products.

SOURCES OF IONISING RADIATION (*ANNEX II*)

Foodstuffs may be treated only by the following sources of ionising radiation:

(a) gamma rays from radionuclides 60 Co or 137 Cs;

(b) X-rays generated from machine sources operated at or below a nominal energy (maximum quantum energy) level of 5 MeV;

(c) electrons generated from machine sources operated at or below a nominal energy (maximum quantum energy) level of 10 MeV.

POINTS OF MAJOR IMPORTANCE

The following points, relating to issues of political importance, became the subject of lengthy discussions. The last point is only mentioned, because it has a particular interest for third countries exporting irradiated foodstuffs to the EU.

- the establishment of the Community list of irradiated products in stages (Art. 4.2),
- the conditions under which existing national authorisations would operate in a transitional period (Article 4.4).
- the compulsory labelling of any irradiated food and food ingredient, even when incorporated in compound ingredients used in finished products (Art. 6.1c),
- the official approval of irradiation installations of any country (Art. 7 and 9)

For the completion of the list the Eur. Commission will make a proposal to the EP and the Council by the end of year 2000 (Art. 4.3), taking into account existing national authorisations and the opinions of the Scientific Committee for Food. Member States have already sent their suggestions to the Commission. The decision on their eventual transfer to the Community list will be taken by codecision.

National authorisations existing at the time of adoption of the directives may be maintained until the entry into force of the completed list foreseen by Art. 4.3, provided they are cleared by the SCF and comply with the framework directive; otherwise they cease to be valid (Art 4.4).

No national authorisation can be issued any more. The restriction of issuing new authorisations does not apply for products that had been authorised in other Member States and cleared by the SCF (Para. 5).

Until the entry into force of the directive completing the Community list, Member States may retain their national bans, *in compliance with the general rules of the EU Treaty* (Para. 7). This clarifying condition is meant to have an important role if a banning measure is challenged before the Court. In such a case, the judge would normally have to examine whether the measure obstructing the free circulation of goods is justified on health grounds and if so, whether it is proportionate.

Irradiated foodstuffs from third countries are subject to the same conditions regulating their treatment, as in the European Union. Their imports may not be permitted, unless they comply with the provisions of Article 9.1 The irradiation facilities carrying out the treatment have to be approved by EU. They must also be subject to an official supervision in their country of origin, guaranteeing that the requirements of Article 7 are respected. The list of irradiation facilities approved will be published in the *Official Journal of the European Communities* (Art. 9.2).

For the purpose of drawing up this list, the Commission may instruct experts to carry out, under its authority, evaluations and inspections of irradiation facilities in third countries, in accordance with Article 5 of Directive 93/99/EEC, on the *additional measures concerning the official control of foodstuffs*[1]. The Commission may conclude technical arrangements with the competent organisations in third countries on the procedures whereby the evaluations and inspections referred to above are to be carried out.

The preparatory work for the approval procedure is expected to start soon. It is therefore in the interest of countries exporting to the EU products covered by the implementing directive, (*dried aromatic herbs, spices and vegetable seasonings*), or by national authorisations of Member States, to start considering whether their facilities fulfil the conditions of Art. 9.1 of the framework directive. For the approval procedure of Art. 9.2, some guidelines are expected to be issued; however, individual approvals will take place only on request of the irradiation facilities.

[1] OJ L 290, 24. 11. 1993, p.14.

Legal and Technological Consequences of the Ongoing Harmonization of Regulations in the Field of Food Irradiation

D. A. E. EHLERMANN
Institute of Process Engineering
Federal Research Centre for Nutrition
Haid-und-Neu-Str. 9
D-76131 Karlsruhe, Germany

Abstract

This paper attempts to analyse certain legal and technological consequences of the efforts of FAO and IAEA through their Joint FAO/IAEA Division, Vienna and the International Consultative Group on Food Irradiation (ICGFI) in harmonizing regulations on food irradiation in their Member States in the past decade. Mainly, these consequences result from the way how the acceptable dose values are defined and whether such provisions can be enforced in practice. Also operators of irradiation facilities and food manufacturers have to be concerned about the consequences of such regulations as unfounded restrictions will affect the profitability of their services. Harmonization, literally, makes regulations only more homogeneous, but not necessarily more reasonable. In order to contribute to free trade under the WTO-agreement, harmonization would have to remove any incompatible, unfounded restrictions. This is outlined by referencing existing national regulations and contrasting them to recent harmonization efforts.

Introduction

Processing of food by ionizing radiation is a fascinating topic where profound scientific evidence and deeply anchored, public prejudice meet to be converted into intelligent regulations. This not always results in solutions which allow for the exploitation of the full benefits of this technology. Furthermore, all parties to the WTO agreement have assumed the obligation to remove 'technical barriers' from trade ie to rely on sound science in their regulations. However, passing regulations requires to take into account also political developments and especially the growing common concern about industrialization of food production and supply, regardless whether justified or not. This has led to a wide variety of national regulations which hamper the exploitation of this technology within a country but also make nearly impossible international trade. In this context, even the General Standard of Codex Alimentarius on food irradiation [1] can be interpreted in different ways. Consequently, efforts for harmonization have been initiated by the IAEA very early, to provide references for regulations and for their harmonization [2, 3] and to observe and report the actual developments [4]. Governments also showed keen interest in this promising technology [5], some converting the General Standard of Codex Alimentarius directly into national law, others seeking to develop harmonized legislation in concordance with other governments in their region. The crucial quantity in radiation processing is absorbed dose; it determines whether the treatment given is justified to fulfil a technological need or to serve a hygienic purpose, the prerequisites of Codex Alimentarius [1]. It is also this physical quantity dose which is regulated in many different ways.

Philosophy of regulating food processes

There are two fundamental approaches for any law system, because a lack of a specific legal provision may imply

- no prohibition to do or apply

or

- only allowable if explicitly permitted.

As the second alternative would block any development and would make the administrative and legislative procedures very complicated and complex, usually a mixture of both approaches is implemented: in the first step the process of interest is generally banned and then only specific exemptions are granted. However, this situation necessarily causes that the kind of interest in and especially any concerns about the process at the time the interest or concern is raised determines how the process be regulated and how difficult it becomes to obtain such individual, explicit permissions. In the case of processing food by ionizing radiation, the justification for introducing regulations and by this way restrictions for the application was that not enough was known about the effects of this technology. 'Not enough' must be read as that decades ago there were founded reservations based on sound science. Logically this would imply, that as soon as such reservations are removed on the evidence again of sound science also the restrictions implemented should be lifted. In reality, this will not always occur because of the nature of such political processes. Consequently, the existing laws in several countries differ because of the different reasons which led to the first step to regulate food irradiation at all and then to the subsequent steps of introducing individual provisions to lift the restrictions partially. Up-to-date, only Singapore was keen enough to consider the provisions of the Codex Alimentarius alone as sufficient for a national regulation. This is being followed by several other countries in Asia especially those from ASEAN [5].

The terminology of Codex Alimentarius

The first standard of Codex Alimentarius on food irradiation in 1979 relied on provisional conclusions of the JECFI (Joint Expert Committee on Food Irradiation) in the years before and granted several conditional clearances for individual applications and at specific maximum dose values. However, when the 1980 JECFI [6] found irradiated food safe to eat regardless of kind of the food and purpose of the irradiation treatment, still a number for the maximum tolerable dose was given. This physical quantity was termed 'overall average dose' and its sole value was 10 kGy. The idea behind these decisions was that in the dose range of interest any radiation-chemical effect was linearly proportional to radiation absorbed dose and, hence, a hypothetical toxicological risk would not depend on a single ingestion but on the longer-term consumption of a range of food treated at a variety of doses. 'Overall average dose' was not intended to cover the technological validity of this 10 kGy-value. For example, potatoes irradiated at 10 kGy are completely unpalatable, have lost their sensory and technological properties, however, remain non-toxic and are still safe to eat. Of course, this was a theoretical approach which was based on sound scientific evidence. Logically, this first revision of the General Standard of Codex Alimentarius [1] now gave up the concept of individual regulations, adopted instead the concept of 'overall average dose' and the sole value of 10 kGy and granted an unconditional clearance for the process. At that time, it could not be foreseen whether such rules and definitions would be suitable in practice for regulation and control of the commercial utilization of food irradiation. Today, it can be seen that the terminology from toxicology implemented in the General Standard is misinterpreted as a quantity suitable for regulations and their enforcement by control and inspection.

The technological terminology in food irradiation

In principle, the technology of processing food by ionizing radiation is 'self-regulating'. ie, there is no need to set administrative regulations for the doses applied. A too low dose would result in a failure to achieve the desired effect; a too high dose - more than needed to achieve the desired effect - would be a waste of energy and a sacrifice of costs, and a far too high dose would even result in an impairment of the quality of the product and a total loss. From the technological viewpoint there is no real concern about the set maximum tolerable dose as the majority of applications is below the present limiting value of 10 kGy from the present Codex Alimentarius; and the few possible high-dose applications would be of no concern considering the 'overall' principle. As an example, the US regulation for spices allowing a maximum dose of 30 kGy was explicitly founded on the reasoning [7] that a very low amount of the food intake irradiated at somewhat higher doses could not introduce any toxicological hazard. But this regulation took into account a minimum dose for spices of about 8 kGy, that some irradiation facilities operate at a maximum-to-minimum-dose ratio of up to 3, consequently at a maximum dose of about 24 kGy, which including a margin for rounding values is just 30 kGy (!) as regulated. This is an exemplary model how technological conditions should influence legislative procedures; however, in practice things are mostly the other way round.

The ongoing harmonization

The most effective world-wide harmonization in the area of radiation processing of food would be to accept the General Standard of Codex Alimentarius [1] as the sole reference. This is also the only acceptable approach under the WTO agreement. However, through the efforts of FAO, IAEA and ICGFI, recently a number countries have already implemented or are implementing a somewhat more general approach (ASEAN countries, Mexico, Pakistan, Ghana and Turkey for example) declaring the General Standard as the basic reference but specifying at the same time food groups or classes, application purposes and limiting dose values. In this situation and with the view on facilitation of international trade in irradiated food, IAEA with the help of several governments initiated harmonization workshops for the Asia/Pacific region (1993), for African countries (1996), for Latin America and the Caribbean (1997) and for the Near East (1998). All these workshops ended in draft model regulations converting the provisions of the General Standard of Codex Alimentarius into more practicable regulations. In the light of the most recent developments and the results of the Joint Study Group on High-Dose Irradiation [8] the latest workshop recommended 'advisory technological dose limits' [9] as the sole reference for regulations. This proposal removes the unfortunate clause of 'overall average dose' and leaves the decision open whether in regulations specific figures for dose values are necessary at all. It is the most prudent starting-point for the ongoing harmonization [5]. However, it is not yet incorporated in a revision of the Codex General Standard. And it is only the ASEAN countries which have initiated the procedures to convert such models in a co-ordinated effort into harmonized regulations of its member states.

In comparison, the new EU-Directives on food irradiation [10, 11, 12] also serve literally the purpose of harmonization of the laws of the Member States by allowing only for the minimum common to all Member States. Reducing for all Member States the extent of dearances for radiation processing to nearly zero, in some sense, can also be termed 'harmonization'. The directives also fall back to the use of 'overall average dose' which was confounded from the beginning [13] when the EU prepared for this harmonization. Unfortunately, the provisions now introduced do not improve the situation and do not contribute to facilitating international trade. Before, the EU-members such as United Kingdom and The Netherlands had already re-drafted and improved their legislations in order to resolve in particular the problem of dose limits [14, 15]; now, this solution will be annulled and superseded by the restrictive EU-Directive.

Nothing to do with harmonization has the situation that many countries implement in their regulations provisions which allow for imports of irradiated food under the condition that the national conditions are met by the treatment abroad. A harmonization worth this name would have to acknowledge in such importation clauses any treatment which is legal in the country of origin. This can be based on the assumption that any country permitting radiation processing of food has done so on the foundation of sound science, which will not differ by nations.

Legal aspects

The General Standard of Codex Alimentarius does not restrict the utilization of radiation processing to a group or class of food and obviously not to specific foodstuffs [1]. Its Annex B gives only 'examples' for which the utility had been shown. Consequently, any regulation limiting the process to certain foodstuffs is already at variance with the General Standard. Furthermore, in its Annex A on dosimetry under no. 1 the General Standard gives an explanation of 'overall average dose'; however, in practice this quantity cannot be measured and, hence, be inspected and controlled; it is also of no practical relevance in radiation processing (with rare exceptions) and, hence, not monitored at the irradiation facility. It is therefore a challenge to authorities to introduce also for radiation processing of food a legislation which can be enforced [16]. Such prudent approaches would be to follow the results of all these world-wide harmonization efforts [5] and to regulate only 'advisory dose limits' and to group the list of such values by classes of food and classes of applications.

Over-regulation negatively affects the profitableness of the technology without a justification based on sound science. Also the case of labelling must be considered in this context. There is a fundamental right of consumers to know [17] which needs to be converted into practical regulations. That is to say, a food labelled for being irradiated must be treated within the 'advisory technological dose limits' and the information from the label must be traceable back to the irradiation facility and the records kept there. There is no chance to determine the exact dose using analytical methods on the food under suspicion [18]. This problem to control the correct dose will become unsurmountable in case the labelling of ingredients for being irradiated is required regardless of the amount of a component. At present, the General Standard [1] together with the

Labelling Standard [19] requires such labelling without exception. The EU-Directive [10] likewise has such requirement; and the EU is also blocking any modification of the respective Codex Standard. The task for an inspector now would be, once he finds on the market a product labelled as "pizza (several main ingredients), with salami containing pepper (irradiated)" to collect the pepper kernels, to identify the irradiated ones, to measure their dose and to estimate the 'overall average dose'. It is also essential for a consumer wishing to buy an irradiated product to be assured that is was irradiated in fact. If the inspector cannot find a single pepper kernel treated at the appropriate dose the labelling 'pepper (irradiated)' would be false and termed fraudulent! Consequently the manufacturer of such pizza would have to strive that at least one slice of salami on each pizza carries at least one cut through an irradiated pepper kernel. Such strict requirements would make impossible the exploitation of radiation processing of ingredients; such labelling would also be misleading to the consumer which is also against some basic Codex rules.

Technological aspects

An operator of a facility for radiation processing is only concerned about the efficacy of his services; that, generally, is to apply the required and contracted minimum treatment which is necessary to achieve the intended purpose and not to ruin the product by an excessive treatment or to lose energy and throughput (that is money) by administering an unnecessary treatment. This obvious rule also applies to processing food by ionizing radiation and 'treatment' is absorbed dose which is also the regulated quantity. Only a legislation relying on the concept of 'advisory technological dose limits' provides to the processor a range of doses for the treatment which allows for the necessary flexibility. The terminology of 'advisory' implies that the utility of the treatment has been generally acknowledged at those dose values; in case a need occurs to depart from such dose values it implies only that such need should be substantiated.

As the operator of an irradiation facility exclusively strives to meet such minimum and maximum dose values, quite automatically records are collected about the doses administered. This is an information which is most suitable for inspection and enforcement of a correct radiation treatment. The adherence to such minimum and maximum technological dose limits must not be hampered by too restrictive regulations.

Over all, there are also technological reasons to harmonize the legal provisions. As an example, radiation processing of spices aims at the efficient elimination of microorganisms. As discussed above, the required minimum dose is about 8 kGy and in some irradiation facilities the resulting maximum dose would be about three times the minimum dose, in our example about 25 kGy. Whereas the US regulation [7] prudently allows for such treatment, the EU-members United Kingdom and The Netherlands indirectly allow for a maximum dose of 15 kGy, France has regulated a maximum value of 11 kGy. In the first case the minimum effective dose of 8 kGy has as a consequence a maximum to minimum ratio of about two which can be met at most irradiation facilities; the latter case with a resulting maximum to minimum ratio of about 1.4 can be met at several irradiation facilities with extra efforts to reduce the dose span, eg by partial loading of the carriers. However, there is no sound reason based on scientific evidence for such restrictions. Furthermore, if spices have to be irradiated to fulfil all these diverging regulations this would severely hamper international trade and make control of the correct treatment extremely exhausting.

Outlook

The WTO-agreement has become effective and a number of cases have been submitted to its conciliation procedures. Processing of food by ionizing radiation may become another case. However, some regional efforts and achievements in harmonizing the relevant regulations and restricting them at the same time to the minimum necessary and based on sound, scientific science leaves some hope for the future. Obviously, ASEAN member states have already taken the lead implementing reasonable regulations. Other countries are not too far from such approach; only the recent EU-Directives fell back into a status which is historically explainable but fails to acknowledge recent developments elsewhere.

References

[1] Codex General Standard for Irradiated Foods and Recommended International Code of Practice for the Operation of Radiation Facilities used for the treatment of foods, FAO/WHO, Rome, 1984, CAC/VOL. XV - Ed. 1

[2] Anon., Preparing Regulations for the Control of Food Irradiation Facilities (ICGFI Document No. 1), IAEA, Vienna, 1991.

[3] Anon., Guidelines for the Authorization of Food Irradiation Generally or by Classes of Food (ICGFI Document No. 15). IAEA, Vienna, 1991.

[4] Anon., Control of Irradiated Food in Trade - a compilation of principles and international recommendations for regulatory control measures (ICGFI Document No. 21), IAEA, Vienna, 1995

[5] P. Roberts, this meeting

[6] Wholesomeness of food, Report of a Joint FAO/IAEA/WHO Expert Committee, WHO, Geneva, 1981, Techn. Report Series 659

[7] Anon., Irradiation in the Production, Processing, and Handling of Food (21 CFR Part 179), Fed. Reg. 51 (1986) no.75, 376-399

[8] Anon., Food Safety - Joint FAO/IAEA/WHO Study Group on High Dose Irradiation, Weekly Epidemiological Record 73(1998) no.3, 9-11 (full report in press)

[9] Report of the Joint AAEA/FAO/IAEA Regional Workshop on "Present status and guidelines for preparing harmonized legislation on food irradiation in the Near East", Tunis, Tunisia, 12 16 October 1998

[10] Directive 1999/2/EC of the European Parliament and of the Council of 22 February on the approximation of the laws of the Member States concerning foods and food ingredients treated with ionising radiation, Official Journal of the EC (.3.1999), L66/16 -L66/23

[11] Directive 1999/3/EC of the European Parliament and of the Council of 22 February on the establishment of a Community list of foods and food ingredients treated with ionising radiation, Official Journal of the EC (13.3.1999), L66/24 - L66/25

[12] H. Vounakis, this meeting

[13] Reports of the Scientific Committee for Food (Eighteenth Series), Commission of the European Communities, Luxembourg, 1987

[14] The Food (Control of Irradiation) Regulations 1990, Statutory Instruments 1990, no. 2490, 1-15

[15] Besluit von 2 april 1992, houdende Warenwetbesluit Doorstraalde waren, Staatsblad van het Koninkrijk der Niederlanden (1992) 205, 1-21
[16] D.A.E. Ehlermann, Control of food irradiation: A challenge to authorities, Trends Food Sci. Technol. 4 (1993), 184-189
[17] Anon., Guidelines for Consumer Protection, United Nations, New York, 1986, ST/ESA/170
[18] Ehlermann, D.A.E., Dosimetry and identification as a tool for official control of food irradiation, Radiat. Phys. Chem. 46 (1995), 1242-1245
[19] Codex General Standard for the Labelling of Pre-Packed Foods, FAO/WHO, Rome, 1991, Codex STAN 1-1985, Rev.1

Irradiation as an Alternative Treatment to Methyl Bromide for Insect Control

S. C. GUPTA
The United States Department of Agriculture
Animal and Plant Health Inspection Service
4700 River Road, Unit 133
Riverdale, MD 20737, USA

ABSTRACT

Methyl bromide is currently in widespread use as a fumigant. The environmental effects of using methyl bromide are being scrutinized by international, Federal, and State agencies. The United States Environmental Protection Agency (EPA), based on its evaluation of data concerning the ozone depletion potential of methyl bromide, published regulations in the Federal Register on December 10, 1993 (58 FR 65018-65082). That rule froze methyl bromide production in the United States at 1991 levels and required the phasing out of domestic use of methyl bromide by the year 2001. The EPA revised the accelerated phase-out regulations that govern the production, import, export, transformation and destruction of substances that deplete the ozone layer under authority of Title VI of the Clean Air Act Amendments of 1990 (CAA) and established a 25 percent reduction in the 1991 baseline levels of production allowances and consumption allowances for methyl bromide (class I, Group VI controlled substance) for the 1999 and 2000 control periods (Federal Register, June 1, 1999, 64 FR 29240-29245). This rule also allows the EPA to amend these regulations through notice and comment rule-making for a complete phase-out of production and consumption with processes for special exemptions permitted under the Montreal Protocol by the beginning of January 1, 2005. The EPA plans to publish a proposal that will describe a process for exempting quantities of methyl bromide used in the U.S. for quarantine and pre-shipment from the reduction steps in the phase-out schedule.

The Agriculture, Rural Development, Food and Drug Administration, and Related Agencies Appropriations Act, 1999 provides a quarantine-use exemption for the production, importation, consumption of methyl bromide to fumigate commodities entering or leaving the United States for purposes of complying with APHIS regulations. The EPA has also indicated that it will work closely with USDA, State agricultural departments, and other stakeholders to define the pre-shipment and quarantine uses of methyl bromide that will be exempt from the phase-out. The EPA's proposal assumes the continued availability of methyl bromide for use as a fumigant for at least the next few years. Nonetheless, the Animal and Plant Health Inspection Service (APHIS) of the United States Department of Agriculture (USDA) is studying the effectiveness and environmental acceptability of alternative treatments to prepare for the eventual unavailability of methyl bromide fumigation.

The APHIS has been exploring alternatives to methyl bromide to remain in compliance with the regulations promulgated by the EPA. Researchers with the Agriculture Research Service of USDA and several academic institutions around the world are actively engaged in finding solutions to combat agriculture pests. Radiation treatment of many traded commodities represents a viable alternative to methyl bromide fumigation and has been found to be one of the best quarantine treatments in reducing and combating the entry of harmful pests into the United States. Scientists are actively investigating the tolerance doses for a number of tropical fruits, and the minimum doses for several pests of quarantine importance. This paper discusses the status, benefits and risks of irradiation over methyl bromide fumigation for controlling insects that infest plant and products.

INTRODUCTION

Food safety is a subject of paramount interest and concern not only in the United States but around the world today. In the United States alone, researchers estimate that up to 33 million Americans, roughly 14 % of its population, become ill from microorganisms in their food each year. Microbial food borne disease is the number one food safety hazard with estimates of $4.5 billion impact in medical costs and lost productivity in the United States each year. The use of irradiation can dramatically reduce food-related disease, destroy insects and other organisms in produce, lengthen shelf life of fruits and vegetables and offers alternative to the use of certain fumigants and chemicals as plant quarantine treatments. These facts, coupled with the proposed cancellation of the preferred fumigant, methyl bromide, indicate food irradiation will have an increasingly important role in the future. At least 42 countries have approved irradiation treatment of more than 50 foods or groups of related foods. It is important to mention that more than 30 countries are using irradiation technology to combat food borne pests. The total quantity of food irradiated in 1997 is about 201,800 tons worldwide as reported on the International Consultative Group on Food Irradiation website (http://www.iaea.org/icgfi/). With regard to phytosanitary irradiation treatments to control plant pests, the International Plant Protection Convention, and the North American Plant Protection Organization have endorsed the technology as effective and safe.

The Animal and Plant Health Inspection Service (APHIS), USDA is aware of growing commercial interest in the use of irradiation as a treatment for agricultural products, both for food safety purposes (to kill pathogens and retard spoilage) and for phytosanitary purposes (to destroy plant pests). In anticipation of requests to allow the use of irradiation in APHIS' regulatory programs, we have been developing policies for evaluating irradiation methods and have been evaluating research on the efficacy of irradiation. We published a notice entitled, "The Application of Irradiation to Phytosanitary Problems" in the Federal Register on May 15, 1996 (61 FR 24433-24439) and received several pertinent comments. This notice discussed how APHIS, in collaboration with the Agricultural Research Service (ARS), would evaluate scientific research to determine the minimum irradiation doses necessary to kill or render sterile particular pests associated with particular articles. The notice emphasized that minimum dose levels are important and necessary, but that dose levels by themselves do not constitute a complete treatment schedule or an adequate regulatory framework. Treatment schedules, in addition to specifying minimum doses, may employ irradiation as a single treatment, as part of a multiple treatment, or as a component of a 'systems approach' combined with other pest mitigation measures. The regulatory framework for employing irradiation treatments must also address system integrity or quality control issues, including methods to ensure that the irradiation is properly conducted so that the specified dose is achieved. These guidelines must also address matters such as packaging or safeguarding of the treated articles to prevent re-infestation. I plan to discuss highlights of the APHIS proposal entitled, "Irradiation Phytosanitary Treatment of Imported Fruits and Vegetables" which is under development.

IRRADIATION ALTERNATIVE TO FUMIGATION BY METHYL BROMIDE

Irradiation is an efficacious and versatile process to use as a quarantine treatment to kill or render sterile a particular pest associated with a particular article. Packages of foods to be treated in a commercial irradiator usually should take less than 30 min to receive a low dose of up to 1 kGy. This process is much faster than fumigation of fruits which usually takes 2 hr plus one hr for aeration. Similarly, thermal treatment, the vapor heat process requires $4^1/_2$ to 5 hr of treatment, and one hr for cooling (Moy and Hara 1994). Irradiation has distinct advantages as a quarantine treatment such as its application to products that are already packaged, a low damage incidence to many agricultural commodities, and speed of treatment as indicated above (Hallman and Worley, 1999 and references therein). It has few disadvantages to consider such as it does not kill insects in a reasonable time at the doses allowed on fresh commodities by the U.S. Food and Drug Administration (< or = 1 kGy). Higher doses of ionizing radiation can be detrimental to the quality of fruits and vegetables.

THE APHIS has been proactive in establishing policy and standards for the use of irradiation to mitigate pests of plants and plant products. Detail citations are listed below for your information:

- A proposal on the "Irradiation phytosanitary treatment of fresh fruits and vegetables" is being developed in APHIS and will be published in the near future.
- Fruit from Hawaii. 63 FR 65645-65649, November 30, 1998. Incorporated into the Code of Federal Regulations at 7CFR 318.13-4f.
- Irradiation in the production, processing and handling of food. 62 FR 64107-64121, December 3, 1997. Incorporated into the Code of Federal Regulations at 21CFR 179.
- Irradiation in the production, processing, and handling of animal food and pet food. 60 FR 50099, September 28, 1995. Incorporated into the Code of Federal Regulations at 21CFR 579.
- Mediterranean fruit fly; additions to quarantined areas and treatments. (Interim rule). 62 FR 47553-47558, September 10, 1997. Incorporated into the Code of Federal Regulations at 7CFR 301.78-10.
- Papaya, carambola, and litchi from Hawaii. 62 FR 36967-36976, July 10, 1997. Incorporated into the Code of Federal Regulations at 7CFR 318.13-4f.
- The application of irradiation to phytosanitary problems. (Notice of policy). 61 FR 24433-24439, May 15, 1996
- The use of irradiation as a quarantine treatment for fresh fruits of papaya from Hawaii. 54 FR 387-393, January 6, 1989. Incorporated into the Code of Federal Regulations at 7CFR 318.13-4f.

After early discussions at the NAPPO annual meetings in 1994, APHIS issued a clear policy on irradiation as a phytosanitary treatment in May 1996 (APHIS, 1996).

EFFICACY OF IRRADIATION QUARANTINE TREATMENTS

As discussed above, irradiation has been demonstrated to be effective in killing or devitalizing organisms that may contaminate and harm commodities or the ecosystems to which the commodities move. Complete mortality or sterility of insects is the main objective of all quarantine treatments. The measures of efficacy of irradiation disinfection treatments should be prevention of adult emergence, when only eggs and larvae are present or sterility when pupae or adult are present. Pest groups such as tephritid fruit flies, curculionid weevils, and some Homoptera do not require high doses of irradiation and hence low doses of irradiation can be tolerated by many fruits. It has been found that female insects are more susceptible to radiation-induced sterility than males however some spider mite (Tetranychidae) females are more tolerant than males (Hallman, 1998). Irradiation doses needed to cause rapid mortality reduce the quality of fresh commodities. For example, it has shown that sweetpotato weevil, *Cylas formicarious elegantulus* (Summers), adult irradiated with 1kGy lived up to 10 days (Dawes *et al.*, 1987) and this dose is sufficient for this pest based on its mortality but sweetpotatoes cannot tolerate this dose without significantly affecting the quality of the cooked roots (McGuire and Sharp 1995). Hallman (1998) also discusses in his paper that when developing quarantine treatment, one should concentrate on the most tolerant stage of the insect present in the commodity for example to achieve 100% mortality of flight-able adult of cowpea weevils it requires a dose of >2.0 kGy of radiation as compared to 20 Gy of radiation for egg stages (data summarized in Table 1, Diop *et al.*, 1997).

Table 1. Doses to achieve 100% adult mortality of cowpea weevils, *Callosobruchus maculatus* (F.), irradiated at different life stages (after Diop *et al.*, 1997).

Stage	Dose (Gy)
Egg	20
7 day-old larva	20
15 dat-old larva	30
21 day-old larva	60
Pupa	140
Flightless adult	1,500[1]
Flight-able adult	> 2,000[1]

[1] Mortality within 48 hr.

Table 2 provides doses which might achieve quarantine security against several groups of pests with a word of caution that this needs further research to satisfy the degree of confidence by taking a larger sample of these insects (Table from Hallman, 1998). It is clear from these observations that more research is needed to establish irradiation doses for additional insects.

The purpose of these treatments is to minimize the pest risk and to maximize the safety associated with the movement and use of commodity. Treatments and associated procedures are based on science, and are no more restrictive than necessary to protect agricultural health.

Table 2. Absorbed dose ranges which might achieve quarantine security of several groups (from Hallman, 1998).

Pest group	Objective	Dose (kGy)
Bruchid weevils	Sterilize adult	0.07-0.1
Aphids, whiteflies	Sterilize adult	0.05-0.1
Tephritid flies	Prevent adult emergence from third instar	0.05-0.25
Scarab beetles	Sterilize adult	0.05-0.15
Curculionid Weevils	Sterilize adult	0.1-0.2
Noctuidae and Tortricidae	Prevent adult emergence from third instar	0.1-0.3
Pyralidae and Tortricidae	Sterilize late pupa	0.2-0.3
Tetranychid mites	Sterilize adult	~0.3
Stored product beetles	Sterilize adult	0.05-0.4
Acarid mites	Sterilize adult	~0.5
Stored product moths	Sterilize adult	0.1-1
Root-knot nematodes	Sterilize adult	~4.0

APHIS'S CURRENT PROPOSAL ON USE OF IRRADIATION AS A QUARANTINE TREATMENT FOR FRUITS AND VEGETABLES

In the following sections, I plan to present APHIS's current thinking based on available literature and recommendations by various international organizations. This proposal discusses following issues and how they must be integrated to achieve effective irradiation treatments, and then proposes specific standards for an irradiation treatment for fruit flies in imported fruits and vegetables.

I. Critical Control Points: Dose, Dosimetry, Safeguards
II. Irradiation Doses to Control Fruit Flies in Fruits and Vegetables
III. Dosimetry and Dose Control Issues
IV. Safeguards for Different Irradiation Situations

I. Critical Control Points: Dose, Dosimetry, Safeguards

Three critical control points are identified in the activities involved in irradiating imported fruits and vegetables to prevent the spread of fruit flies. These are points where errors will definitely reduce the long-term effectiveness of the treatment and where, on the other hand, correct procedure will ensure effective treatments.

The three critical control points are:

Dose: The dose of ionizing radiation, calculated in Gray, must be sufficient to prevent adult emergence of each species of fruit fly in fruits and vegetables. Each dose is set at the lowest level that achieves this effect; the dose will not necessarily kill larvae immediately after treatment. These doses are based on research conducted by ARS and others, as discussed below..

It is important that the dose be set at the lowest effective level for regulatory, economic, and product quality reasons. The Food and Drug Administration has issued regulations providing that fruits and vegetables may receive up to 1 kGy of irradiation (21 CFR 179.26). This current limit of 1 kGy for fruits and vegetables is significant because industry irradiation methods can only ensure that all articles in an irradiated lot receive a guaranteed minimum dose, at the cost of having some articles in the lot subjected to two or three times the minimum dose. Therefore, to achieve a minimum absorbed dose of 250 Gray, some articles in a lot may be subjected to a dose of 750 Gray or more. This avoids the possibility of any articles being subjected to a dose above 1 kGy. Also, the higher the dose, the greater the cost of the irradiation treatment. Finally, irradiation causes many fruits and vegetables to suffer changes in color and texture at higher doses.

Dosimetry: If establishing the required dose correctly is the first critical control point, delivering the expected dose accurately and consistently is the second critical control point. Accurate dosimetry ensures that this happens. An effective dosimetry system is necessary to ensure that irradiated articles do in fact receive the minimum required dose of ionizing radiation. An inaccurate dosimetry system that records received doses as higher than they actually are, could allow survival of fruit flies in treated articles. An inaccurate dosimetry system that records received doses as lower than they actually are, could result in doses exceeding the 1 kGy limit, as well as unacceptable changes in the color and texture of the fruits and vegetables.

Safeguards: The third critical control point, safeguards, addresses the movement and identification of articles before and after they are irradiated. There is always a risk that treated articles may become re-infested with pests after treatment, and safeguards are necessary to control this risk. If the fruits and vegetables are irradiated after arriving in the United States, safeguards must also be employed to ensure that pests do not escape from articles en route through the United States to the irradiation facility. Finally, systemic safeguards, such as record-keeping, labeling, and monitoring and enforcement of regulatory requirements, are necessary to ensure that articles are not accidentally or intentionally presented as properly irradiated when they have not been.

II. Irradiation Doses to Control Fruit Flies in Imported Fruits and Vegetables

APHIS is proposing regulations providing for the use of irradiation as a phytosanitary treatment to control 10 species of fruit flies in imported fruits and vegetables (Table 3).

Table 3. Irradiation doses for 10 species of fruit flies with their common and scientific names.

Fruit Flies in Imported Fruits and Vegetables		
Scientific Name	Common Name	Dose (Gray)
Bactrocera dorsalis	Oriental fruit fly	250
Ceratitis capitata	Mediterranean fruit fly	225
Bactrocera cucurbitae	Melon fly	210
Anastrepha suspensa	Caribbean fruit fly	150
Anastrepha ludens	Mexican fruit fly	150
Anastrepha obliqua	West Indian fruit fly	150
Anastrepha serpentina	Sapote fruit fly	150
Bactrocera tryoni	Queensland fruit fly	150
Bactrocera jarvisi	(No common name)	150
Bactrocera latifrons	Malaysian fruit fly	150

The recommended doses are sufficient to ensure probit 9 efficacy (a statistical estimation of 99.9968 percent mortality or sterility, corresponding to a survival rate of 32 fertile flies per million). The doses will almost entirely prevent emergence of live adults from irradiated fruits and vegetables.

ARS found sufficient data in its review of research to recommend irradiation doses for 10 species of tephritid fruit flies.

In 1986, minimum doses of 150 and 300 Gray were internationally proposed for quarantine security of tephritid fruit flies and all other arthropods, respectively. It was concluded that the dose of 150 Gray should prevent development of adult tephritid fruit flies capable of flight when eggs and larvae are irradiated, while 300 Gray should cause sterility to all stages of other insects and mites.

APHIS has proposed irradiation doses based on the pertinent literature. The doses vary according to species, because the resistance of species varies. We propose to require minimum doses of 250 Gray and 225 Gray for *B. dorsalis* and *C. capitata* based on the study of Seo *et al.* (1973). A minimum dose of 210 Gray is proposed for *B. cucurbitae* because at a dose of 209 Gray no *B. cucurbitae* larvae of an estimated 169,903 in bell peppers reached the adult stage (Seo *et al.* (1973). APHIS proposes a minimum dose of 150 Gray for seven other tephritid fruit flies: *B. tryoni, B. jarvisi, B. latifrons, A. suspensa, A. serpentina, A. ludens,* and *A. obliqua.* Although the research evidence shows that lower doses might suffice, a dose of 150 Gray should pose no greater problem to irradiation-tolerant commodities compared with lower doses, and it provides greater security.

III. Dosimetry and Dose Control Issues

It is critical to ensure that articles actually receive the required dose during irradiation, since lower doses could allow pests to survive. Fortunately, the irradiation industry and researchers have spent decades developing and documenting effective systems for dosimetry. We are confident that facilities that correctly apply dosimetry guidance published by the American Society for Testing and Materials (ASTM) will be able to reliably measure the doses regulated articles receive.

The basic product delivered by businesses engaged in irradiation is an accurately measured dose of ionizing radiation, within the range requested by the customer, delivered to articles provided by the customer. Therefore, dosimetry is an integral part of these businesses' procedures, and APHIS recommends that businesses follow good dosimetry practices to map, control, and record the radiation dose. Guidance and requirements for dosimetry by ASTM and the U.S. Department of Energy, and supervision by responsible national agencies in foreign countries should be used to establish the degree of dosimetry reliability.

IV. Safeguards for Different Irradiation Situations

Safeguards reduce risk by controlling the movement and identification of articles before and after they are irradiated. Safeguards include such matters as packaging, labeling, records, and irradiation facility construction and procedures. The types of safeguards needed to reduce risks will change with changing conditions. Certain safeguards are needed if the irradiation facility is located in an area infested with fruit flies, where the risk is high that flies could oviposit in fruit after it is irradiated. These identical safeguards would not be needed if the irradiation takes place in an area not infested with fruit flies. The goal of the safeguards is to address risks that are not fully addressed by the technical irradiation components of this proposal -- delivering an effective dose of radiation and monitoring this delivery through accurate dosimetry. Such risks include misidentification of articles so that untreated cartons are delivered labeled as treated, re-infestation of treated articles after treatment, and escape of fruit flies from articles in the United States prior to treatment.

In order to be certified, a facility would have to be capable of administering the minimum absorbed ionizing radiation doses and be constructed according to specified standards so that treated and untreated fruits and vegetables are kept in separate locations with a physical barrier between them to prevent the transfer of cartons. Treated and untreated fruits and vegetables could be separated with barriers such as a 6-foot wall or chain link fence to minimize interference with facility operations and visibility.

It might seem that an insect proof barrier should separate the areas for untreated and treated articles. Obviously, a chain link barrier would not prevent flies from emerging from fruit on the untreated side, flying to the treated side, and leaving the facility in fruit that has been irradiated. However, APHIS believes that an insect proof barrier in the facility is unnecessary.

An irradiation facility should be located in the northern states of the United States or overseas in a country where the climate does not allow fruit fly outbreaks to become established, or in a region that has been determined to be free from fruit flies.

A facility in the United States would only be approved to irradiate imported regulated articles if APHIS Management determines that regulated articles could be safely transported to the facility from the port of arrival without significant risk that plant pests will escape in transit or while the regulated articles are at the facility. The compliance agreement for a facility located in the United States would require the facility to comply with additional requirements to prevent escape of plant pests from the articles prior to their treatment. One of these requirements would be prompt irradiation of the fruits and vegetables to minimize the risk that fruit flies could emerge from the articles and either spread from the facility or (as discussed above) spread to treated articles and reinfest them.

Treatment would have to be monitored by an APHIS inspector, who would also inspect treatment records and make unannounced inspections of the facility. APHIS has proposed that facilities which operate continuously should notify an inspector at least 24 hr before the date operations commence. Facilities that carry out periodic irradiation operations would have to notify an inspector at least 24 hr before scheduled operations. Monitoring and verification are extremely important to ensure the integrity of the entire system for irradiating imported fruits and vegetables. This is because there is no practical way for an inspector to determine, based on physical evidence from the commodity itself, that a commodity has been irradiated. Irradiation leaves no residue and usually causes no discernable change to the commodity's color or texture. In addition, an effective irradiation treatment may not kill all larvae, but instead might prevent adult emergence.

Fruits and vegetables irradiated before arrival in the United States would have to be packaged to prevent re-infestation after treatment. Articles to be irradiated in facilities in the United States will generally be imported in shipping containers that are not opened until they reach the facility. Therefore, there are no packing requirements for articles that are to be irradiated in the United States, although importers must sign compliance agreements to ensure against possible releases of fruit flies en route. Such articles may be packed in ventilated cartons that are not insect-proof. Packing requirements are not needed in this case because the articles are destined for irradiation in northern States where fruit flies cannot become established and spread. Any fruit flies that might escape in the vicinity of the facility are, therefore, not a risk, and any fruit flies associated with the articles will be irradiated and are, therefore, not a risk.

We propose to require that absorbed dose be measured at the treatment facility using a dosimetry system that can accurately measure a minimum absorbed dose of 150, 210, 225, and 250 Gray, to match the required doses for the relevant species of fruit fly. We would require that the dosimetry system, including the number and placement of dosimeters used at the facility to measure the absorbed dose, be in accordance with standards of ASTM.

We considered proposing to require that each carton or other container of fruits or vegetables contain a radiation indicator that is placed in the carton or other container prior to irradiation and remains in place until after the fruits and vegetables have entered the United States. Such indicators, in the form of chemically impregnated stickers that change color after exposure to specified levels of ionizing radiation, are inexpensive and are in common use on cartons of medical supplies irradiated for sterility.

The treatment facility must maintain records of treatment for 1 year after each lot is treated. APHIS would require the records to include the lot identification, scheduled process, evidence of compliance with the scheduled process, ionizing energy source, source calibration, dosimetry, dose distribution in the product, and the date of irradiation. All of these records should be kept by irradiation facilities, either as normal business practice or as required by the Nuclear Regulatory Commission or State agencies.

From the literature available, we believe the fruits and vegetables authorized for treatment under this section are tolerant to the minimum absorbed dose required by the treatment. However, the facility operator and shipper would be responsible for determination of tolerance.

Damage to treated commodities is a significant issue for importers. In fact, one reason there is interest in irradiation treatments is that the treatment is effective against fruit flies in a number of tropical fruits and other articles that do not survive other treatments well; e.g., papaya, rambutan, carambola, litchi. Some research has shown changes to fruit color and texture when they are irradiated at several times the required minimum dosage. This could be a problem because some irradiation facilities may subject the articles to two to three times the required minimum dose -- either to ensure that all articles in a treated lot (both those nearest to and farthest from the radiation source) receive at least the required minimum dose, or because it is simply infeasible to expose the articles to the radiation source and then remove them quickly enough to achieve only the minimum required dose.

CONCLUSION

From the above account, it is clear that irradiation is one of the most useful and effective quarantine treatments to replace the fumigant, methyl bromide. Methyl bromide is known to damage the ozone layer and the under the Montreal Protocol, the EPA in the United States is required to phase-out the use of this chemical. The EPA recent regulations (EPA, 1999) require a reduction of 25% of methyl bromide by year 2000 and complete phase-out by year 2005 but only by amending the most recent regulations. This rule also provides for some flexibility for imports of fruits and vegetables. Under the Montreal Protocol and Clean Air Act, the developing countries have up to year 2015 to implement the phase-out of methyl bromide. This should give sufficient time for these nations to develop systems to control the use of ozone depleting chemicals such as methyl bromide for quarantine purposes. The Animal and Plant Health Inspection Service of USDA is committed to make changes in its regulatory framework to accommodate the recommendations made by International Treaties such as Montreal Protocol and to comply with regulations established by the EPA. There is no doubt that irradiation is an excellent choice as a quarantine treatment but this is an expensive undertaking for many developing countries in Latin America and Asia. I recommend that efforts should be maintained in harmonization of irradiation regulations in the developed nations, and education and capacity building for the developing nations. It is of paramount importance that extensive research be supported for determining irradiation doses for different stages of insects of quarantine importance keeping in mind that such treatments do not cause spoilage of fruits and vegetables.

ACKNOWLEDGMENTS

I am pleased to acknowledge the assistance of Mr. Peter Grosser and Mr. Richard Kelly for providing the necessary background information on this subject. I am grateful to Ms. Carolyn Cohen for critically reading the manuscript and providing useful comments.

REFERENCES

[1] Animal and Plant Health Inspection Service, United States Department of Agriculture: 1993. Importation of Unmanufactured Wood Articles From Mexico: Proposed rule: 7 CFR Part 319.

[2] Environmental Protection Agency (EPA): June 1, 1999. Protection of Stratospheric Ozone: Incorporation of Montreal Protocol Adjustment for a 1999 Interim Reduction in Class I, Group VI Controlled Substances, Federal Register 64(104): 29240-29245.

[3] Animal and Plant Health Inspection Service, United States Department of Agriculture: 1996. The application of irradiation to phytosanitary problems (Notice of Policy), Federal Register 61(95): 24433-24439.

[4] Dawes, M.A., Saini, R.S., Mullen, M.A., Brower, J.H., and Loretan, P.A. 1987. Sensitivity of sweetpotato weevil (Coleoptera: Curculionidae) to gamma irradiation, J. Economic Entomology, 80: 142-146.

[5] Diop, Y.M., Marchioni, E., Ba, D., and Hasselmann, C. 1997. Radiation disinfestation of cowpea seeds contaminated by *Callosobruchus maculatus*, J. Food Process. Preserv. 21: 69-81.

[6] Food and Drug Administration, 1986. Irradiation in the production, processing, and handling of Food. Federal Register 51(75): 13376-13399.

[7] Hallman, G.J. 1998. Ionizing radiation quarantine treatments. An. Soc. Entomol. Brasil, 27(3): 313-323.

[8] Hallman, G.J. and Worley, J.W. 1999. Gamma radiation doses to prevent adult emergence from immatures of Mexican and West Indian Fruit Flies (Diptera: Tephritidae), J. Economic Entomology, 92(4): 967-973.

[9] Moy, J.H. and Hara, A.H. 1994. The efficacy and versatility of irradiation as a quarantine treatment. NAPPO proceedings

[10] Satin, M. and Loaharanu, P. 1997. Irradiation as an alternative post harvest treatment. Proceedings of the Seminaro Nacional Aceptacion Y Comercio De Alimentos Irradiados, October 27-29, 1997, Toluca, Mexico.

Epidemiology of Foodborne Illness—Role of Food Irradiation

R. A. ETZEL
Director, Division of Epidemiology and Risk Assessment
Office of Public Health and Science, Food Safety and Inspection Service
United States Department of Agriculture
Washington, DC, USA

ABSTRACT

Foodborne illness is a serious problem throughout the world. In the United States, the Centers for Disease Control and Prevention estimates that 76 million illnesses and almost 5,000 deaths per year can be attributed to food-related hazards. U.S. estimates of the annual cost of medical treatment and lost productivity vary widely, from $6.6 billion to $37.1 billion from the seven major foodborne pathogens. Only a fraction of the people who experience foodborne illness seek medical care. In those who do seek care, and who submit stool specimens, bacterial agents are most often determined to be the cause of the illness. The bacterial agents most often identified in patients with foodborne illness in the United States are *Campylobacter, Salmonella*, and *Shigella* species, although substantial variation occurs by geographic area and season. Although testing for viral etiologies of diarrheal disease is rarely done, viruses are considered to be the most common cause of foodborne illness in the United States. Irradiation of foods, approved by the U.S. Food and Drug Administration in 1997 and by the U.S. Department of Agriculture's Food Safety and Inspection Service in 1999 for use on raw meat and meat products, has the potential to reduce the risk of foodborne illness in the United States.

Introduction

Foodborne illness has been a serious problem in the United States and other countries for centuries. During the early 20th century, typhoid fever affected about 1 out of every hundred people in the United States. By 1920 the rate decreased to 33.8 per 100,000 and by 1950 to 1.7 per 100,000 (1). Many factors contributed to the decline in typhoid fever, including the passage of the Pure Food and Drug Act (2). In the early 21st century, we continue to be faced with challenges in making the United States food supply safer. With approval of food irradiation in 1997 by the U.S. Food and Drug Administration and by the U.S. Department of Agriculture's Food Safety and Inspection Service in 1999 for use on raw meat and meat products, we have at hand the opportunity for the use of this technology to decrease microbial pathogens on foods.

The goals of this manuscript are to:
Describe the foodborne diseases surveillance system in the United States
Summarize new estimates of the burden of foodborne illness in the United States
Describe the potential impact of food irradiation on the rate of foodborne illness in the United States, and
Emphasize the importance of continued surveillance to help evaluate new food safety programs, including food irradiation.

In this manuscript, it is not possible to give estimates of the potential impact of food irradiation on foodborne illness rates in other countries because the etiology of foodborne illness may be quite different in other countries than in the United States.

United States Surveillance for Foodborne Diseases

Like many countries, the United States has a difficult time accurately assessing the scope of the problem of foodborne illness. To improve surveillance, in 1995 the Food Safety and Inspection Service (FSIS) began a collaborative project with the Centers for Disease Control and Prevention (CDC), the Food and Drug Administration (FDA), and five state health department sites to collect more precise information on the incidence of foodborne disease in the United States. This project is known as the Foodborne Diseases Active Surveillance Network, or FoodNet (3).

An active surveillance system for foodborne diseases was needed to get a better estimate of the burden of foodborne illness in the United States. Prior to 1999, the best estimate was that there were between 6.5 and 33 million cases of foodborne disease in the United States each year, with up to 9,000 deaths (4,5). United States estimates of the annual cost of medical treatment and lost productivity vary widely, from $6.6 billion to $37.1 billion dollars from the seven major foodborne pathogens (6). More precise estimates of morbidity and mortality from foodborne diseases were needed than could be made with the limited data from existing passive surveillance systems.

In the United States, most of what we know about foodborne disease comes from outbreak investigations. The majority of cases of foodborne diseases, however, are sporadic. Most of these sporadic cases are not captured by passive surveillance systems, such as the national laboratory based surveillance system, because of the chain of steps required for a case to be included. A person must first become ill and develop symptoms, the symptoms must be severe enough for the person to seek medical care, the physician must order a specimen for testing, the specimen must be collected and submitted to a clinical laboratory, the laboratory must be able to test for the organism and report positive results to local or state public health agencies. Therefore, it is not surprising that in the United States only about 5 percent of foodborne disease cases which occur in the community are reported to public health officials.

To understand the earlier surveillance efforts in the United States, it is important to note that before 1995, there were three foodborne disease surveillance systems in place at many state health departments and the Centers for Disease Control and Prevention. The Foodborne Outbreak Disease Surveillance System is a database of foodborne outbreaks (defined as two or more cases of a similar illness resulting from ingestion of a common food) investigated by state and local health departments and reported to CDC. Each year only about 500 outbreaks and 20,000 cases are reported to CDC through this surveillance system. In addition to information from outbreaks, all 50 state public health department laboratories report culture-confirmed cases electronically to CDC. A third source of data are epidemic investigations conducted by CDC in collaboration with State and local health departments. About 12-16 outbreaks are investigated by CDC each year. These three surveillance systems are all considered passive systems, since they rely on reporting from either local or state health departments or clinical microbiology laboratories. The missing source of data for foodborne disease was active surveillance, and that is why FoodNet was established.

In 1999 there are active FoodNet sites in Connecticut, Minnesota, Oregon, and selected counties in California, Georgia, New York, Maryland and Tennessee (6). The total population in these catchment areas in 1999 is about 7.5% of the United States population.

With FoodNet, active surveillance is done for 7 bacterial pathogens (*Campylobacter,* Shiga toxin-producing *E. coli* O157, *Listeria, Salmonella, Shigella, Vibrio, Yersinia*) and 2 parasites (*Cryptosporidium* and *Cyclospora*). All clinical microbiology labs within each state's catchment area are contacted by the state either weekly or monthly, depending on the size of the clinical laboratory, to ensure complete case ascertainment.

FoodNet also collects data on the outcome of patients with foodborne disease. Hospitalization rates were highest for persons with *Listeria* infection. *Listeria* and *Salmonella* caused most of the reported deaths.

Additional studies of laboratories, physicians, and the general population were conducted in the FoodNet sites to evaluate the factors that influence surveillance. From the population survey, it was determined that Americans have about 1.4 diarrheal episodes per person per year. Only about 8% of persons with diarrheal illness sought medical care for their illness. About 20% of these persons provided a stool specimen for culture.

The United States also has a system called PulseNet, a national computerized database of molecular fingerprints of pathogens which allows scientists to quickly compare strains of bacteria to determine whether or not there is an outbreak and also to determine if there may be a common source.

In December of 1998, FoodNet and PulseNet enabled the US government to quickly identify a multistate outbreak of listeriosis caused by an unusual strain of *Listeria monocytogenes* (7). The outbreak involved 100 illnesses and 21 deaths in 22 states. Fifteen of the deaths were among adults, and 6 were fetal deaths. Because of the Pulsenet system, CDC was able to link several patients in different states who all had a very rare strain of *Listeria*, known as Type E on Pulsed Field Gel Electrophoresis. The CDC and the Food Safety and Inspection System worked together to identify the plant which had produced the hot dogs and luncheon meats that patients had consumed. Samples of unopened meat products from that plant contained the same Type E *Listeria*, prompting a large recall of the contaminated products. This rapid public health response would not have been possible before FoodNet and PulseNet were in place.

Estimates of the burden of foodborne illness in the United States

New estimates of the public health burden of foodborne disease in the United States were released by CDC in 1999 (9). To show how FoodNet data were used to estimate the burden of foodborne diseases, nontyphoidal *Salmonella* will be used as an example. Through active surveillance, an average of 2,090 culture-confirmed *Salmonella* cases per year were reported in 1996 and 1997. Since FoodNet catchment sites at that time represented about 5.6% of the United States population, this translates into an estimated 37,171 cases in the United States. CDC scientists then multiplied by a factor of 38 (to account for underreporting) to estimate the total number of 1,412,498 cases of nontyphoidal *Salmonella* in the U.S. (9).

The CDC estimates indicate that in the United States 76 million cases of foodborne illness occur yearly, with about 5,000 deaths (9). The known pathogens account for 14 million illnesses per year, and the unknown agents account for 62 million illnesses per year. An estimated 325,000 hospitalizations for foodborne illnesses occur yearly. Known pathogens account for 60,000 hospitalizations per year, and unknown agents account for 265,000 hospitalizations per year. Of the estimated 5,000 deaths each year, known pathogens account for 1,800 deaths and unknown agents account for 3,200 deaths. Among all illnesses attributable to foodborne transmission, about 30% are caused by bacteria, 3% are caused by parasites, and 67% are caused by viruses (9).

Potential impact of food irradiation on the rate of foodborne illness in the United States

Irradiation of foods has been approved by the U.S. Food and Drug Administration and by the U.S. Department of Agriculture's Food Safety and Inspection Service. If widely adopted, it has the potential to reduce the risk of foodborne illness in the United States.

What proportion of foodborne illness might be eliminated in the United States if food irradiation was maximally utilized? If 33% of illnesses attributable to foodborne transmission are caused by bacteria and parasites, then an estimated 1/3 of foodborne illnesses (those attributable to bacteria or parasites) may have the potential for prevention by food irradiation. Foodborne illnesses caused by viruses are not likely to be preventable by food irradiation.

Food irradiation technology could be useful to reduce the burden of illness from *Campylobacter jejuni*, the most common of all foodborne infections in the United States, with an estimated 2,453,926 cases per year. Current US control measures reduce but do not eliminate the risk. Irradiating poultry has the potential to reduce the burden of disease from *Campylobacter*.

For foodborne disease from *E. coli* O157, about 73,480 cases are diagnosed in the United States each year. Untreatable infection can lead to severe complications, chronic renal failure, and death. Even low doses can cause severe illness. Existing US control measures reduce, but do not eliminate the risk.

An estimated 2518 cases of listeriosis occur each year in the United States. About 20% of pateints die of meningitis or bacteremia. At particularly high risk are the elderly, pregnant women, and those with compromised immune systems. High risk products include processed meats and soft cheeses, among other foods. Epidemics have resulted from in-plant recontamination after an initial kill step.

It is clear that irradiation should not be a substitute for improved sanitation on the farm and in slaughter or processing plants. But given its potential to reduce the bacteria and parasites on foods, it may be helpful in reducing foodborne illnesses in the United States.

Importance of continued surveillance

Active foodborne disease surveillance systems are useful, not only to estimate the burden of foodborne diseases in humans, but also to evaluate the effectiveness of intervention programs, such as food irradiation. Analysis of temporal and geographic trends in surveillance data is vitally important to evaluate the effectiveness of newly-approved interventions. Evaluating surveillance data may help to answer the question "Did the intervention help to reduce foodborne diseases in humans?"

Conclusion

The Foodborne Diseases Active Surveillance Network (FoodNet) is supported by a strong and productive collaboration of eight state health departments, CDC, FSIS and FDA. By rapidly receiving and reporting cases of foodborne disease, this collaboration is providing data necessary for monitoring the burden of foodborne diseases and evaluating new interventions.

References

[1] CDC. Safer and healthier foods. MMWR 1999;48:905-913.
[2] Young JH. Pure food: securing the Federal Food and Drugs Act of 1906. Princeton, New Jersey:Princeton University Press, 1989.CDC
[3] CDC. The Foodborne Diseases Active Surveillance Network, 1996. MMWR 1997;46:258-61.
[4] Archer DL, Kvenberg JE. Incidence and cost of foodborne diarrheal disease in the United States. J Food Prot 1985;48:887-894.
[5] Council for Agricultural Science and Technology. Foodborne pathogens: risks and consequences. Council for Agricultural Science and Technology, Ames, Iowa, 1994.
[6] Buzby JC, Roberts T. Guillan-Barre syndrome increases foodborne disease costs. Food Rev 1997;20:36-42.
[7] CDC. Incidence of foodborne illnesses: preliminary data from the Foodborne Disease Active Surveillance Network (FoodNet) –United States, 1998. MMWR 1999;48:189-194.
[8] CDC. Update: Multistate outbreak of listeriosis—United States, 1998-1999. MMWR 1999;47:1117-1118.
[9] Mead PS, Slutsker L, Dietz V, McCaig LF, Bresee JS, Shapiro C, Griffin PM, Tauxe RV. Food-related illness and death in the United States. Emerging Infect Dis [serial online] September-October 5 (5).
Available from http://www.cdc.gov/ncidod/eid/vol5no5/mead.htm

Irradiation: A Critical Control Point in Ensuring the Microbiological Safety of Raw Foods

R. A. MOLINS
Food and Nutrition Board
Institute of Medicine
National Academy of Sciences
Washington, DC, USA

Y. MOTARJEMI
Food Safety Unit
Programme of Food Safety and Food Aid
World Health Organization
Geneva, Switzerland

F. K. KÄFERSTEIN
Food and Drug Administration and
Food Safety and Inspection Service
Washington, DC, USA

ABSTRACT

The widespread and increasing incidence of foodborne diseases and the resultant social and economic impact on the human population have brought food safety to the forefront of public health concerns. This has prompted public health authorities worldwide to reassess their methods of food safety assurance, and to resort to a more cost-effective, preventive method that is known as Hazard Analysis and Critical Control Point (HACCP).

Ensuring food safety depends on effective control measures, i.e. methods to prevent food contamination and, when necessary, decontamination. Present production methods cannot totally prevent food contamination, and the complexity of food handling and processing provides ample opportunity for contamination as well as survival and growth of pathogenic organisms.

The application of an HACCP-based approach as a method for the management of hazards of the food chain demonstrates the need for applying a cold decontamination treatment as a control measure in the production of foods which are to be marketed raw or minimally processed. Irradiation (referred to as cold pasteurization) is such a control measure in the production of several types of raw or minimally-processed foods such as poultry, meat and meat products, fish, seafood, and fruits and vegetables. In the production of these foodstuffs, irradiation is thus a Critical Control Point. It has the potential to eliminate vegetative forms of bacterial pathogens as well as parasites. Moreover, irradiation fulfills other criteria for a Critical Control Point, i.e. critical limits (minimum and maximum doses) can be established and monitored, and process control is well known. Corrective actions can also be taken when necessary.

Irradiation is a safe technology and has been recognized as such by the FAO/WHO Codex Alimentarius Commission. It certainly merits the attention of industry and public health authorities. Today, 40 countries permit the irradiation of one or more foodstuffs: 12 countries have approved its use for pathogen control in poultry, 7 other for use in meats and 13 in fish and seafood.

I. INTRODUCTION

A Joint FAO/WHO Expert Committee on Food Safety concluded as early as 1983 that "illness due to contaminated food is perhaps the most widespread health problem in the contemporary world and an important cause of reduced economic productivity" (WHO, 1984).

Data published since then by various countries confirm this statement and indicate that that the problem may be on the increase (Sockett, 1993). Statistics from industrialized countries show that up to 10% of the population in these countries may suffer annually from a foodborne illness (WHO, in press). Although reliable statistics on foodborne diseases are not available from most developing countries because of a lack of surveillance systems (Motarjemi and Käferstein, 1997; Todd, 1997a), the high incidence of diarrhoeal diseases in these countries is an indication of the gravity of the situation. Worldwide, over 4,000 million episodes of diarrhoea occur annually (WHO, 1997a). The great majority of these cases occur in the developing world.

In countries like the United States, a great proportion of foodborne illnesses are due to diseases such as salmonellosis and campylobacteriosis (Buzby and Roberts, 1997). In contrast, developing countries are subject to a wider range of foodborne illnesses, including a large number of parasitical diseases (Todd, 1997b). In addition, public health authorities all over the world are being challenged by emerging, re-emerging or newly-recognized types of foodborne illnesses such as verocytotoxin-producing *Escherichia coli* (VTEC) infections, listeriosis and other bacterial infections (Bennett, 1996; Damgaard "et al.," 1996; Johnson "et al.," 1996; Wesley, 1996), infections by protozoa such as *Cyclospora cayetanensis* (Jackson "et al.," 1997), and foodborne trematode infections, often with severe, chronic or fatal health consequences. The magnitude of the foodborne illness problem and its health and economic consequences have been well documented in the literature (Sockett, 1993; Buzby and Roberts, 1997; Motarjemi and Käferstein, 1997). These reviews testify to the importance of these diseases for health and the economy, and the need for their prevention.

Contamination of foods with pathogenic organisms also is playing an increasingly important role in the economy of many countries and may even have devastating effects on their foreign trade, as happened in Peru during the initial stages of the cholera epidemic in the early 1990´s (Petrera and Montoya, 1992). In international trade, rejections of food shipments or even total bans on food imports from specific countries or regions are frequently imposed by importing countries on the basis of microbial contaminants, sometimes without regard to scientific merit (WHO, 1995a; Gilbert, 1996; Merican, 1996; Buzby and Roberts, 1997).

On the other hand, microbial contamination of foods may cause severe economic losses to processing companies or entire industries even when the hazard is recognized before consumers are affected. Recalls of foods because of microbial contamination accounted for 25% of all food recalls in the United States between October, 1991 and September, 1992 (Venugopal "et al.," 1996). A recent, massive recall of some 10,000 tonnes of ground beef in the U.S.A. because of contamination with *E. coli* O157, which has captured headlines throughout the world, will likely result in losses in the millions from destroyed products, as well as from recall, disposal and decreased consumer confidence.

The health and economic consequences of foodborne illnesses, combined with decreasing resources to combat them, have led both public health authorities and food industries to recognize weaknesses in traditional food safety assurance methods and the need to resort to a more efficient, preventive method known as Hazard Analysis Critical Control Point, HACCP, originally developed by the Pillsbury Company (1973). The application of HACCP in some countries has since then become mandatory for certain food industries (USDA, 1996). Furthermore, HACCP is expected to play an important role in future trade in food commodities throughout the world (Adams, 1994; Merican, 1996). The main reason for this is the relevance given the Codex Alimentarius by the Agreement on the Application of Sanitary and Phytosanitary Measures (SPS) of the World Trade Organization, which has made Codex its basic food safety reference body (Wehr, 1996). Based on this development, important trading countries are striving to adapt their food safety standards so that those of the Codex Alimentarius can be formally adopted.

The HACCP system can be applied to the entire food chain as a food safety management tool for the identification of measures which are necessary for the prevention of foodborne illnesses. At the processing level, on the other hand, and properly applied, HACCP may offer the best means to enhance the safety of fresh produce or prepared foods by reducing the incidence and levels of pathogenic organisms (Anon., 1991; Bauman, 1995). Application of the HACCP system allows identification of hazards and assessment of the risk of such hazards (Anon., 1993), hopefully leading to the establishment of control measures which are essential for achieving food safety .

This article aims to demonstrate that, by adopting a HACCP-based approach to risk management of raw foods, particularly solid foods of animal origin, some fresh produce and other foods that are marketed either raw, under-cooked or minimally processed, the application of a decontamination treatment such as irradiation becomes a necessary control measure in ensuring their safety and thus, a Critical Control Point (CCP). Failure to apply this technology as a corrective intervention measure means that the safety of many such foodstuffs cannot be guaranteed presently through HACCP because the assumption that the only other critical control point, heating, will be properly applied by the food preparer is faulty. In addition, irradiation is a control measure which fulfills the conditions for a critical control point in that a) it eliminates or reduces to acceptable levels the risk posed by pathogenic bacteria and parasites in these foods; b) critical limits are well established for this processing technique, and c) compliance with critical limits can be accurately monitored through dosimetry.

The focus of the paper will be on bacterial and parasitical agents, as irradiation at the doses required for cold pasteurization of raw foods does not eliminate viruses from the food (Grecz "et al.,", 1983).

II. HAZARD ANALYSIS OF THE FOOD CHAIN

Food production today, particularly in urban or industrialized settings, is a complex chain of events. From the point of production to that of consumption foodstuffs go through a number of stages and operations, each a potential source of contamination and an opportunity also for survival and growth of microorganisms. In line with the HACCP approach, the hazards at each step in the food chain are identified, and possible control measures suggested.

Production of agriculture/aquaculture products

A large number of studies show that raw foodstuffs are frequently contaminated with one or several types of foodborne pathogens. In many industrialized countries, a great percentage of poultry products, up to 60% or higher, has been shown to be contaminated with pathogens such as *Salmonella* and *Campylobacter* (Roberts, 1990; D'Aoust, 1995). For instance, in the United Kingdom, despite reductions in the level of contamination, 33 to 41% of UK-produced retail raw chicken in 1993 was found to be contaminated with *Salmonella* or *Campylobacter* (Anon., 1994; Jacob, 1995). Raw meat products and fish as well as fresh produce may also be contaminated with pathogens such as *Salmonella* (Beuchat, 1995), *Yersinia* (Khare "et al.," 1996), *E. coli* O157 (Feng, 1995; IFST, 1996), *Listeria monocytogenes* (Hudson and Mead, 1989; Rocourt, 1994) or with various parasites (Eckert, 1996).

The source of the contamination is diverse. Depending on the food it can vary from methods of animal rearing, including contamination of feeds and environment, mass rearing, automatized slaughtering (Bell, 1997), use of contaminated or waste water for agri- or aquaculture, use of manure as fertilizers, conditions of harvesting and processing (Brocklehurst "et al.," 1987), and contamination of the environment. These diverse factors may lead to contamination of food with various pathogens of bacterial, parasitical or viral nature. Although good agriculture/aquaculture practices can significantly reduce the level of microbial contamination of foods, it is an established fact that presently, even under the best conditions of production, this contamination cannot be totally prevented. Raw foods or ingredients represented the causative factor of nearly 23% of 766 outbreaks of foodborne illness in the United States between 1977 and 1982 (Bryan, 1989). Some countries also lack the resources to implement measures to minimize food contamination. In addition, it should be realized that some of the pathogens are part of the environment and will inevitably be present in the food chain.

Therefore, at production level, there is no control measure to prevent microbial contamination to a degree that foodstuffs can be considered as free from pathogens. The fact that raw milk may be contaminated is recognized worldwide, and for this reason, there is a mandatory requirement for milk pasteurization in many countries.

Processing, transportation and distribution

Many food processing technologies based on heat treatment, such as canning and pasteurization, effectively control microbial hazards in foods and render them safe if applied properly. However, by definition, raw foods cannot be subjected to bactericidal heat treatment; similarly, some minimally-processed foods may not undergo any heating (Fain, 1996). Although many other treatments, especially chemical treatments such as washing fresh produce with chlorinated water (Beuchat, 1992), spraying beef carcasses with organic acid solutions (Dickson and Siragusa, 1994), and immersing poultry carcasses in phosphate solutions (Rodriguez de Ledesma et al., 1996), effectively decrease the numbers of pathogenic bacteria in the final products, none can ensure elimination of bacterial hazards, let alone parasites. Consequently, pathogens may survive in foodstuffs which are either minimally-processed or marketed in the raw state.

During transportation and distribution there are additional opportunities for growth of microorganisms which are present in the food, and for cross-contamination between foodstuffs. Temperature abuse of foods that should be kept at refrigeration temperatures is a common occurrence leading to increased risks from foodborne pathogens (Palumbo, 1986; Bryan, 1989;); this factor is particularly important in countries having deficiencies in the cold chain or in which electric power failures are frequent.

Preparation and serving

The preparation step is very important in the food chain as it is the last step before consumption. Any hazard which is not controlled at this step is likely to lead to disease. Epidemiological data from foodborne disease surveillance have demonstrated that most foodborne disease episodes occur as a result of one or a combination of several errors during food preparation, among others (i) inadequate cooking, (ii) cross-contamination from raw to ready-to-eat food, and (iii) growth of surviving microorganisms as a result of inadequate cooling or storage.

A recent review on the origin of foodborne disease outbreaks points to improper hygienic, food preparation and handling practices in the home as very important factors leading to foodborne illnesses (Scott, 1996). Nevertheless, the major source of reported foodborne disease outbreaks is the institutional kitchen (Bryan, 1981a, b). A major outbreak of *Escherichia coli* O157:H7 in the United States in 1993 was attributed to improperly cooked hamburguer meat (Marks and Roberts, 1993); this outbreak highlighted the frequent failure of institutional food handlers to properly apply the final, corrective heat treatment that would eliminate most microbial hazards (ie., the critical control point). Although food handlers, whether in the home or institutional food serving operations, must be educated in safe food handling, including application of measures which are necessary for rendering food safe (i.e decontamination), experience has shown that this is a formidable task and requires continuous vigilance on the part of the food handlers. The slightest error in food handling can lead to serious health consequences.

In HACCP terms, the above situation in many homes and institutional food preparing and/or serving operations is one in which the Critical Control Point, heating during meal preparation or for serving to temperatures high enough to eliminate the microbial hazards or reduce them to acceptable levels, is left to be applied by often uninformed individuals who, more often than not, lack any means of monitoring the control measure and/or ignore the critical limits. It is equivalent to accepting that the world was as well off when consumers heated the milk at home than with present day pasteurization of milk. Under such conditions, cooking/heating cannot be considered a Critical Control Point (Bryan, 1996), and the conclusion must be reached that HACCP is not being applied to these products. Furthermore, the mere introduction of pathogens in the kitchen through contaminated raw foods that may in turn contaminate surfaces, utensils and the hands of operators is itself a hazard that may lead to cross-contamination of other foods or to recontamination of those already cooked; it is also a disservice to consumers.

Consumption

In addition to the above hazards, some consumers may have a predilection for consuming raw or under-cooked foods, despite the risks (Loaharanu and Sommani, 1991). Although consumers, particularly vulnerable groups, need to be educated about avoiding such foods, it should be recognized that it is very difficult to change the customs and cultures of a population, and before attempting to do so it is worthwhile to consider other more cost-effective interventions.

III. IRRADIATION : A CRITICAL CONTROL POINT IN THE FOOD CHAIN

Educating primary producers in good agri- and aquaculture practices and consumers in safe food handling is an essential control measure in ensuring food safety and preventing foodborne illnesses. However, as indicated earlier, this alone is not sufficient to prevent foodborne illnesses. The efficacy of such measures would be greatly enhanced if irradiation were introduced in the food chain as an additional control measure which can decontaminate many solid raw foods of animal origin and some fruits, vegetables and minimally-processed foods.

Such a control measure could contribute to prevention of foodborne illnesses by:

a) eliminating pathogens such *as Salmonella, Campylobacter, Yersinia, Listeria, Shigella, Vibrio, E.coli* O157, *Toxoplasma gondii*, trematodes, protozoa, *Trichinella* and *Taenia* from these foods;
b) minimizing the risk of cross-contamination at different stages in the food chain, including processing, transportation and distribution or preparation in the kitchen. As the technology can be applied to foods packaged in final retail form, re-contamination during transportation and distribution can also be prevented;
c) minimizing the risk of survival or growth of bacterial pathogens as a result of under-cooking, inadequate cooling or temperature abuse of the food during storage, transportation and distribution, and
d) minimizing the risk posed by foods which are consumed raw or processed using non-bactericidal treatments.

Within the above context, irradiation fully qualifies as a Critical Control Point, defined as "a point, step or procedure at which control can be applied and a food safety hazard can be prevented, eliminated or reduced to acceptable levels" (National Advisory Committee on Microbiological Criteria for Foods, 1995). At present, only irradiation qualifies as a safe, proven and cost-effective, non-thermal treatment to eliminate or reduce pathogenic bacteria to acceptable levels in raw or minimally-processed foods (Thayer "et al.," 1996; Loaharanu, 1996). For such products, irradiation is an essential control measure and one that in the HACCP terminology is referred to as Critical Control Point (CCP) (Molins, 1996a; 1996b).

IV. APPLICATIONS OF IRRADIATION AS A CONTROL MEASURE IN RENDERING FOOD SAFE

Food irradiation is not a new technology. The lethal effects of ionizing radiation on microbes were observed and reported as far back as 1898, and proposals to use radiation to kill bacteria in food were advanced as early as 1916. The amount of research data developed by individual countries and by joint international programmes during the past 50 years surpasses that developed for any other food processing technique. In 1981, a Joint FAO/IAEA/WHO Expert Committee on Food Irradiation (JECFI) gave foods irradiated up to an overall average dose of 10 kGy an unconditional bill of safety: "Irradiation of any food commodity up to an overall average dose of 10 kGy causes no toxicological hazards; hence, toxicological testing of food so treated is no longer required." Furthermore, the JECFI also declared that irradiation of food up to a dose of 10 kGy introduced no special microbiological or nutritional problems (WHO, 1981). Based on these conclusions, in 1984 the Codex Alimentarius Commission issued a General Standard for Irradiated Food and an associated Recommended International Code of Practice for the Operation of Radiation Facilities Used for the Treatment of Foods (Anon., 1984). Many countries followed suit in the 1980s and 1990s and proceeded to enact food irradiation regulations approving the use of this technology for various applications. At present, 40 countries allow irradiation of one or more food products and 29 are applying it commercially (ICGFI, 1997).

Irradiation has many applications in food safety that fully respond to the definition of a Critical Control Point. It is used to eliminate parasites and vegetative forms of bacteria from poultry, meat and meat products, fish and other seafood as well as fruits and vegetables. The technology is sometimes referred to as "cold pasteurization" as it can eliminate pathogens in food without substantial increases in temperature and without causing any significant physical or chemical changes in the food (WHO, 1981; Thayer, 1995; Hampson "et al.," 1996; Thayer "et al.,", 1996; Radomyski "et al.," 1994).

Poultry

The presence of pathogenic bacteria such as *Salmonella* and *Campylobacter* in poultry is a worldwide phenomenon. The contamination problem is aggravated by modern mass rearing practices. Poultry-borne salmonellosis and campylobacteriosis occur mainly as a result of undercooking, time-temperature abuse leading to survival and growth of the pathogens, or recontamination after cooking due to contact with surfaces, hands or utensils which have been previously contaminated with raw chicken. Poultry-borne salmonellosis or campylobacteriosis account for a heavy toll in human disease and associated suffering, including death (Roberts and Unnehever, 1994). High economic costs are also associated with these problems. The United States, for example, reports some 30,000 cases of salmonellosis annually, including more than 3,000 deaths, at a total estimated cost of over 4,000 million dollars. An important proportion of these cases may be due to consumption of poultry. Whether the causes are cross-contamination or under-cooking of poultry, it is clear that if the pathogens are prevented from reaching the homes or institutional kitchens, poultry-borne salmonellosis would be effectively prevented (Scott, 1996).

At the present time, irradiation is the only physical process for rendering contaminated poultry safe other than by heat treatment. A total of 11 countries have so far recognized the importance of this technology for ensuring the safety of poultry products and have approved irradiation of poultry.

Meat and meat products

A similar situation to poultry exists in regard to contamination of meat and meat products, except that the latter may carry a wider range of pathogens including *Salmonella, E. coli* O15 and *Yersinia*, as well as parasites such as *Toxoplasma gondii, Taenia* (*Taenia solium* is particularly dangerous as it may lead to severe neurological disorders in humans), and *Trichinella spiralis*. As for poultry products, under-cooking of meat and meat products often leads to survival of the pathogens (Buchanan, 1986). In the case of time-temperature abuse, bacterial growth may also occur in these products.

An additional problem posed by meat and meat products is that in many countries people enjoy eating them raw or under-cooked. While parasitic agents may be controlled by proper freezing of the meat, for various reasons this is not always practiced, and bacterial pathogens are not eliminated by freezing. As result, meats and meat products which are not irradiated and which are consumed raw or under-cooked are a looming health hazard, as was demonstrated in 1993 by the earlier-mentioned outbreak of foodborne disease caused by under-cooked ground beef contaminated with *E. coli* O157:H7 in the United States; this outbreak affected some 500 persons and caused the death of four children (Marks and Roberts, 1993). It also resulted in a formal petition to the United States Food and Drug Administration to clear the use of irradiation for control of *E. coli* O157 and other pathogenic bacteria in meats (Giddings and Masefield, 1994). This petition was approved in late 1997 as a result of a new incident involving a recall of 10,000 tonnes of ground beef contaminated with *E. coli* O157:H7. Other large outbreaks of *E. coli* O157:H7 in meats and other minimally-processed foods have since occurred in Australia, Japan and Scotland.

Raw meats can be irradiated at relatively low doses to eliminate bacterial pathogens and render parasites uninfective (Lebepe "et al.," 1990; Murano, 1995; Fu "et al.," 1995). Seven countries have approved irradiation of fresh or processed meats for decontamination and/or parasite control.

Fish and Seafood

Fish and seafood are additional groups of food which can be effectively decontaminated through irradiation. Foodborne diseases transmitted by many of these products may include some of those described for poultry and meats (e.g., salmonellosis), as well as diseases caused by aquatic microorganisms, particularly the *Vibrionaceae* (DePaola, A., 1981; Loaharanu, 1996). Among these, *Vibrio cholerae* has been implicated in the cholera pandemic that afflicted many South and Central American countries from 1991 to 1993, with a toll of more than 200,000 cases and thousands of deaths (Albert "et al.," 1997). Other members of this troublesome group of bacteria are *Vibrio vulnificus and Vibrio parahaemolyticus*, frequently found in shellfish.

Irradiation at relatively low doses (1.0-2.0 kGy) effectively eliminates these contaminants (Matches and Liston, 1971) and allows for continued popular enjoyment of such raw fish and seafood dishes as sushi or the popular South American "ceviche", as well as fresh, live oysters and clams. The latter can be radiation-treated without killing the mollusks, thus allowing live storage under refrigerated conditions (Kilgen and Hemard, 1996). Irradiation of fish and seafood is approved in 10 countries and is being considered for approval in several more.

Fruits and vegetables

Fruits and vegetables can be contaminated with various pathogens, including bacteria such as *Salmonella, Shigella, E. coli, Vibrio cholerae, Listeria monocytogenes* or parasites such as *Giardia lamblia, Cryptosporidium parvum, Entamoeba histolytica, Ascaris lumbricoides* and *Cyclospora cayetanensis*. The problem of contamination of fruits and vegetables is more important in developing countries where contaminated water may be used for irrigation or manure used as fertilizer (Beuchat, 1995). However, some industrialized countries such as the USA have experienced an increase in the number of produce-associated foodborne diseases. Increase in international trade in these products may in the future contribute to such an upward trend, as was evidenced by a 1996 outbreak of human infection with *Cyclospora cayetanensis* in the United States attributed to imported blueberries.

Irradiation is an effective control measure for eliminating pathogenic bacteria and parasites from the surface of fruits and vegetables (Langerak and Damen, 1978). Assuming that the large *E. coli* O157 outbreak which occurred in Japan in 1996 was caused by contaminated radish sprouts, as the epidemiological investigations seemed to indicate, it is likely that if these products had been irradiated, the outbreak, which affected some 5700 school children and caused the death of 12, would have been prevented.

V. CRITICAL LIMITS AND MONITORING

In food irradiation processing, critical limits would be the radiation doses necessary to achieve the objective of the treatment, be it elimination of selected pathogenic bacteria or control of parasite infectivity. The radiation doses required to render food microbiologically safe depend on the objectives, i.e. the type of pathogenic hazard to be controlled, the degree of reduction which is sought, and the type of food. Irradiation at medium doses (3.0-7.0 kGy) effectively destroys vegetative cells of all pathogenic bacteria (Thayer "et al.," 1996). Table 1 provides examples of decimal reduction radiation doses (radiation D_{10} values) which have been determined for various non-spore forming, pathogenic bacteria in different foods. Parasites require even lesser doses (< 1.0 kGy), with some exceptions, as described by Loaharanu (1996), since the objective of the irradiation treatment is usually not to destroy the organism but to render it uninfective. Table 2 presents data on minimum effective doses to prevent infection by some important parasitic foodborne pathogens.

Establishment of critical limits in food irradiation for elimination of selected pathogens can readily be accomplished on the basis of D_{10} values reported in the literature. Guidelines on radiation doses required for elimination of bacterial or parasitic hazards in various types of food have been published as Codes of Good Irradiation Practice by the International Consultative Group on Food Irradiation (ICGFI, 1986; 1991a,b) and the IAEA (1993).

For example, inactivation of an unlikely high level of 10^6 colony-forming units (CFU)/g *Escherichia coli* O157:H7 in ground beef (D_{10} = 0.30 kGy) would theoretically require a minimum radiation dose: (6 x D_{10}) = 1.8 kGy as a critical limit. Considering that it is normal to expect a 100-200% difference between the level of radiation absorbed by a food product at the points of maximum and minimum exposure under commercial operating conditions, the ground beef in the example could receive up to 5.4 kGy to ensure that a minimum dose level of 1.8kGy is achieved. As mentioned earlier, application of radiation doses up to10 kGy to food has long been assessed as safe (WHO, 1981; CAST, 1986; WHO, 1994; Diehl and Josephson, 1994; WHO, 1995b), and recognized as such in the Codex General Standard for Irradiated Foods (Anon., 1984). However, a Study Group on High-Dose Irradiation of Foods convened jointly by FAO, IAEA and WHO in September, 1997 concluded that "foods treated with doses greater than 10 kGy can be considered safe and nutritionally adequate when produced under established Good Manufacturing Practice" (WHO, 1997b). It can thus be expected that the Codex General Standard for Irradiated Foods will be ammended in the future to eliminate the present 10-kGy upper limit for overall average absorbed dose.

Process control monitoring in food irradiation, in turn, is well known (ICGFI, 1992). Accurate measurement of the radiation dose absorbed by the product is achieved through standard dosimetry (IAEA, 1977; ASTM, 1996), which should be traceable to national and/or international standards. The Codex General Standard for Irradiated Foods as well as the ICGFI Codes of Good Irradiation Practice have detailed specifications on operation of irradiation facilities, on labelling of irradiated foods, and on the data that should appear in the documentation that must accompany shipments of irradiated foods. Above all, these documents stress the importance of irradiating only high-quality foods and of following at all times Good Manufacturing Practices and the Codes of Hygienic Practices published by the Codex Alimentarius.

VI. CONCLUSION

Present production methods cannot ensure microbiologically safe raw foodstuffs. Although many of these foods can be rendered safe by heat treatment during preparation for consumption, the risk is high that these foods are not decontaminated during food preparation either due to negligence or ignorance. The rising incidence of foodborne illnesses in the world provides the evidence for this. To present consumers or food service operators with raw foodstuffs which may be contaminated and to expect that they render the food safe during its preparation is to transfer the responsibility of safety from producers to food handlers and consumers who, not infrequently, are unaware of the risks posed by the foods and methods of control.

By adopting a HACCP-based approach to risk management, it can be clearly demonstrated that application of a technology such as food irradiation is essential for ensuring the safety of raw foodstuffs, and such an intervention should be considered as a Critical Control Point in the food chain, much as milk pasteurization is today recognized as necessary for ensuring the safety of milk and milk products. Therefore, the potential benefits of irradiation, which is endorsed by national and international bodies such as the World Health Organization, the Food and Agriculture Organization of the United Nations, and others, surely merit serious consideration by public health authorities, industry, and consumer groups worldwide.

VII. REFERENCES

[1] Adams, C. (1994) Microbial concerns of the North and South American countries and scientific implications for harmonizing free trade, Dairy Food Environ. Sanit.14 (8): 471-472.

[2] Albert, M.J., Neira, M., and Motarjemi, Y. (1997) The role of food in the epidemiology of cholera, World Statistics Quarterly 50: 111-118.

[3] Anon. (1984) Codex General Standard for Irradiated Foods and Recommended International Code of Practice for the Operation of Radiation Facilities Used for the Treatment of Food, Food and Agriculture Organization of the United Nations, CAC/Vol XV-Ed.1, FAO, Rome.

[4] Anon. (1991) Draft Principles and Application of the Hazard Analysis Critical Control Point (HACCP) System, Alinorm 93/13, Appendix VI, Codex Alimentarius Commission, Rome.

[5] Anon. (1993) Guidelines for the application of the Hazard Analysis Critical Control Point (HACCP) System, Codex Alimentarius Commission, Joint FAO/WHO Food Standards Programme, FAO, Rome.

[6] Anon. (1994) International Testing, Parallel Food Testing in the UK, EC subsidy: B5-1050/93/0077938, International Consumer Research and Testing Ltd., September.

[7] ASTM (1996) Annual Book of ASTM Standards, ASTM, Philadelphia, PA.

[8] Bauman, H.E. (1995) The origin and concept of HACCP, In: A.M. Pearson and T.R. Dutson (editors), HACCP in Meat, Poultry and Fish Processing, Advances in Meat Research Series, Vol. 10, Blackie Academic and Professional, Glasgow, pp. 1-7.

[9] Bell, R.G. (1997) Distribution and sources of microbial contamination on beef carcasses, J. Appl. Microbiol. 82: 292-300.

[10] Bennett, R.W. (1996) Atypical toxigenic *Staphylococcus* and non-*Staphylococcus aureus* species on the horizon? An Update, J. Food Prot. 59 (10): 1123-1126.

[11] Beuchat, L.R. (1992) Surface disinfection of raw produce, Dairy, Food Environ. Sanit. 12 (1): 6-9.

[12] Beuchat, L.W. (1995) Pathogenic microorganisms associated with fresh produce, J. Food Prot. 59 (2): 204-216.

[13] Brocklehurst, T.F., Zaman-Wong, C.M., and Lund, B.M. (1987) A note on the microbiology of retail packs of prepared salad vegetables, J. Appl. Bacteriol. 63: 409-415.

[14] Bryan, F.L. (1981a) Hazard analysis critical control point approach: Epidemiologic rationale and application to foodservice operations, J. Environ. Health 44: 7-14.

[15] Bryan, F.L. (1981b) Hazard analysis of foodservice operations, Food Technol. 35 (2): 78-87.

[16] Bryan, F.L. (1989) Risks of practices, procedures and processes that lead to outbreaks of foodborne illness, J. Food Prot. 51 (8): 663-673.

[17] Bryan, F.L. (1996) Another decision-tree approach for identification of critical control points, J. Food Prot. 59 (11): 1242-1247.

[18] Buchanan, R.L. (1986) Processed meats as a microbial environment, Food Technol. 40 (4): 134-138

[19] Buzby, J.C. and Roberts, T. (1997) Economic costs and trade impacts of microbial foodborne illness, World Health Statistics Quarterly 50: 57-66.

[20] CAST (1986) Ionizing Energy in Food Processing and Pest Control, I. Wholesomeness of Food Treated with Ionizing Energy, Council for Agricultural Science and Technology Report No. 109, Ames, IA.

[21] Damgaard, P.H., Larsen, H.D., Hansen, B.M., Bresciani, J., and Jorgensen, K. (1996) Enterotoxin-producing strains of *Bacillus thuringiensis* isolated from food, Lett. Appl. Microbiol. 23: 146-150.

[22] D'Aoust, J. (1995) *Salmonella* and the international food trade, Int. J. Food Microbiol. 24: 11-31.

[23] DePaola, A. (1981) *Vibrio cholerae* in marine foods and environmental waters: literature review, J. Food Sci. 46 (1): 66-70.

[24] Dickson, J.S. and Siragusa, G.R. (1994), Survival of*Salmonella typhimurium, Escherichia coli* O157:H7 and *Listeria monocytogenes* during storage of beef sanitized with organic acids, J. Food Safety 14: 313-327.

[25] Diehl, J.F. and Josephson, E.S. (1994) Assessment of wholesomeness of irradiated foods, Acta Alimentaria 23 (2): 195-214.

[26] Eckert, J. (1996) Workshop summary: food safety: meat- and fish-borne zoonoses, Vet. Parasitol. 64: 143-147.

[27] Fain, A.R. (1996) A review of the microbiological safety of fresh salads, Dairy, Food Environ. Sanit. 16 (3): 146-149.

[28] Feng, P. (1995) *Escherichia coli* serotype O157:H7: novel vehicles of infection and emergence of phenotypic variants, Emerging Infectious Disease 1: 47-52.

[29] Fu, A.-H., Sebranek, J.G., and Murano, E.A. (1995) Survival of*Listeria monocytogenes, Yersinia enterocolitica* and *Escherichia coli* O157:H7 and quality changes after irradiation of beef steaks and ground beef, J. Food Sci. 60 (5): 972-977.

[30] Giddings, G. and Masefield, J. (1994) Petition to the Food and Drug Administration: Approved Sources of Ionizing Radiation, Isomedix, Inc., Whippany, NJ.

[31] Gilbert, R.J. (1996) Zero tolerance for *Listeria monocytogenes* in foods - is it necessary or realistic?, Food Australia 48 (4): 169-170.

[32] Grecz, N., Rowley, D.B., and Matsuyama, A. (1983) The action of radiation on bacteria and viruses, In: Josephson, E.S. and Peterson, M.S. (editors), Preservation of Food by Ionizing Radiation, CRC Press, Baton Rouge, pp. 167-218.

[33] Hampson, J.W., Fox, J.B., Lakritz, L., and Thayer, D.W. (1996) Effect of low dose gamma radiation on lipids in five diffrenet meats, Meat Sci. 42 (3): 271-276.

[34] Hudson, W.R and Mead, G.C. (1989) *Listeria* contamination at a poultry processing plant, Lett. Appl. Microbiol. 9: 211-214.

[35] IAEA (1993) Use of Irradiation to Control Infectivity of Food-borne Parasites, Proceedings of a Final Research Co-ordination Meeting, Mexico, 1991, International Atomic Energy Agency, Vienna.

[36] ICGFI (1986) Task Force Meeting on the Use of Irradiation to Ensure Hygienic Quality of Food, International Consultative Group on Food Irradiation, 14-18 July, 1986, Vienna.

[37] ICGFI (1991a) Code of Good Irradiation Practice for Prepackaged Meat and Poultry (to control pathogens and/or extend shelf-life), International Consultative Group on Food Irradiation, ICGFI Document No. 4, Vienna.

[39] ICGFI (1991b) Code of Good Irradiation Practice for the Control of Microflora in Fish, Frog Legs and Shrimps, International Consultative Group on Food Irradiation, ICGFI Document No. 10, Vienna.

[40] ICGFI (1992) Training Manual on Operation of Food Irradiation Facilities, 1st Edition, International Consultative Group on Food Irradiation, ICGFI Document No. 14, Vienna.

[41] ICGFI (1997) Database on food irradiation approvals, Food Preservation Section, Joint FAO/IAEA Division of Nuclear techniques in Food and Africulture, IAEA, Vienna.

[42] IAEA (1977) Manual of Food Irradiation Dosimetry, Technical Report Series No. 178, International Atomic Energy Agency, Vienna.

[43] IAEA (1993) Irradiation of Poultry Meat and its Products - A compilation of technical data for its authorization and control, IAEA TECDOC-688, International Atomic Energy Agency, Vienna.

[44] IFST (1996) Toxin-producing *E. coli* food poisoning and its prevention, IFST Position Paper, International Food Safety News 5 (10): 3-5.

[45] Khare, S.S., Kamat, A.S., Doctor, T.R., and Nair, P.M. (1996) Incidence of *Yersinia enterocolitica* and related species in some fish, meat and meat products in India, J.Sci. Food Agric. 72: 187-195.

[46] Jackson, G.J., Leclerc, J.E., Bier, J.W., and Madden, J.M. (1997) *Cyclospora* - Still another foodborne pathogen, Food Technol. 51 (1): 120.

[47] Jacob, M. (1995) Salmonella in poultry - Is there a solution ?, Environ. Policy and Practice, 5 (2):75-80

[48] Johnson, R.P., Clarke, R.C., Wilson, J.B., Read, S.C., Rahn, K., Renwick, S.A., Sandhu, K.A., Alves, D., Karmali, M.A., Lior, H., McEwen, S.A., Spika, J.S., and Gyles, C.L. (1996) Growing concerns and recent outbreaks involving non-O157:H7 serotypes of verotoxigenic *Escherichia coli*, J. Food Prot. 59 (10): 1112-1122.

[49] Kilgen, M.B. and Hemard, M.T. (1996) Evaluation of commercial irradiation and other processing methods for *Vibrio vulnificus* control in Louisiana oysters, Progress Report, Co-ordinated Research Programme on Irradiation as a Public Health Intervention Measure to Control Foodborne Disease (Cysticercosis/Taeniasis and *Vibrio* Infections) in Latin America and the Caribbean, Joint FAO/IAEA Division of Nuclear Techniques in Food and Agriculture, IAEA, Vienna.

[50] Langerak, D.I. and Damen, G.A.A. (1978) Influence of irradiation on the keeping quality of prepacked soupgreens stored at 10° C, In: Food Preservation by Irradiation Vol. I, International Atomic Energy Agency, Vienna, pp. 275-282.

[51] Lebepe, S., Molins, R.A., Charoen, S.P., Farrar, H., and Skowronski, R.P. (1990) Change in microflora and other characteristics of vacuum-packaged pork loins irradiated at 3.0 kGy, J. Food Sci. 55: 918-924.

[52] Loaharanu, P. (1996) Irradiation as a cold pasteurization process of food, Vet. Parasitol. 64: 71-82.

[53] Loaharanu, P. and Sommani, S. (1991) Preliminary estimates of economic impact of liver fluke infection in Thailand and the feasibility of irradiation as a control measure , Southeast Asian Journal of Tropical Medicine and Public Health 22: 384-390.
[54] Marks, S. and Roberts, T. (1993) *E. coli* O157:H7 ranks as the fourth most costly foodborne disease, Food Rev. 16 (3): 1-8.
[55] Matches, J.R. and Liston, J. (1971) Radiation destruction of *Vibrio parahaemolyticus*, J. Food Sci. 36: 339-344.
[56] Merican, Z. (1996) Dealing with an expanding global food supply, J. Food Prot. 59 (10): 1133-1137.
[57] Molins, R. (1996a) Integration of food irradiation into HACCP plans for minimally processed foods - International regulatory status of food irradiation, EFFoST Conference on Minimal Processing of Food: A Challenge for Quality and Safety, Cologne, Germany, 6-8 November, 1996.
[58] Molins, R. (1996b) Food irradiation: a new critical control point in the context of HACCP approach to food safety, Proceedings, ECA/FAO/IAEA/IIR/ITC/WHO Regional Seminar on Food Irradiation to Control Food Losses and Foodborne Disease, and Facilitate Trade, Rabat, Morocco, 26 February - 1 March, 1996, IAEA, Vienna.
[59] Motarjemi, Y. and Käferstein, F.K. (1997) Global estimation of foodborne diseases, World Health Statistics Quarterly 50: 5-11.
[60] Murano, E.A. (1995) Microbiology of irradiated foods, In: Food Irradiation - A Sourcebook, E.A. Murano (editor), Iowa State University Press, Ames, Iowa, p.42.
[61] National Advisory Committee on Microbiological Criteria for Foods (1995) Hazard Analysis and Critical Control Point System, In: Stevenson, K.E. and Bernard, D.T. (editors), HACCP - Establishing Hazard Analysis Critical Control Point Programs, A Workshop Manual, The Food Processors Institute, Washington, D.C., pp. 2-9.
[62] Palumbo, S.A. (1986) Is refrigeration enough to restrain foodborne pathogens?, J. Food Prot. 49 (12): 1003-1009.
[63] Petrera, M. and Montoya, M. (1992) The economic impact of the cholera epidemic, Peru, 1991, Epidemiological Bulletin, PAHO, 13 (3): 9-12.
[64] Pillsbury Company (1973) Food Safety Through the Hazard Analysis Critical Control Point System, Contract No. FDA 72-59, Research and Development Dept., The Pillsbury Company, Minneapolis, MN.
[65] Radomyski, T., Murano, E.A., and Olson, D.G. (1994) Elimination of pathogens of significance in food by low-dose irradiation; a review, J. Food Prot. 57: 73-86.
[66] Roberts D. (1990) Sources of infection: food, The Lancet 336: 859-861.
[67] Roberts, T. and Murrell, K.D. (1993) Economic losses caused by food-borne parasitic diseases, In: Cost-Benefit Aspects of Food Irradiation Processing, Proceedings of a symposium, Aix-en-Provence, March, 1993, International Atomic Energy Agency, Vienna, pp. 51-57.
[68] Roberts, T. and Unnehever, L. (1994) New approaches to regulate food safety, Food Rev. 17 (2): 2-8.
[69] Rocourt, J. (1994) *Listeria monocytogenes*: the state of the science, Dairy, Food and Environ. Sanit. 14: 70-82.

[70] Rodriguez de Ledesma, A.M., Riemann, H.P., and Farver, T.B. (1996). Short-time treatment with alkali and/or hot water to remove common pathogenic and spoilage bacteria from chicken wing skin. J. Food Prot. 59 (7): 746-750.
[71] Scott, E. (1996) Foodborne disease and other hygiene issues in the home, A review, J. Appl. Bacteriol. 80: 5-9.
[72] Sockett, P.N. (1993) Social and economic aspects of food-borne disease, Food Policy, April 1993: 110-119.
[73] Thayer, D.W. (1995) Use of irradiation to kill enteric pathogens on meat and poultry, J. Food Safety 15: 181-192.
[74] Thayer, D.W., Boyd, G., Fox Jr., J.B., Lakritz, L., and Hampson, J.W. (1995) Variations in radiation sensitivity of foodborne pathogens associated with the suspending meat, J. Food Sci. 60 (1): 63-67.
[75] Thayer, DW, Josephson, E.S., Brynjolfsson, A., and Giddings, G.G. (1996) Radiation pasteurization of food, Issue Paper No. 7, Council for Agricultural Science and Technology (CAST), Ames, Iowa.
[76] Todd, E.C.D. (1997a) Worldwide surveillance of foodborne disease: the need to improve, J. Food Prot. 59 (1): 82-92.
[77] Todd, E.C.D. (1997b) Epidemiology of foodborne diseases: a worldwide review, World Health Statististics Quarterly 50: 30-50.
[78] USDA (1996) Pathogen Reduction; Hazard Analysis and Critical Control Point (HACCP) Systems; Final Rule, United State Department of Agriculture, Food Safety and Inspection Service, 9 CFR Part 304, Federal Register, July 25, 1996, Washington, D.C.
[79] Venugopal, R., Tollefson, L., Hyman, F.N., Timbo, B., Joyce, R.E., and Klontz, K.C. 1996. Recalls of foods and cosmetics by the U.S. Food and Drug Administration. J. Food Prot. 59 (8): 876-880.
[80] Wehr, H.M. (1996) The Codex Alimentarius: Its expanded role in food safety, Dairy, Food Environ. Sanitation 16 (10): 622-626.
[81] Wesley, I.V. (1996) *Helicobacter* and *Arcobacter* species: Risks for Foods and Beverages. J. Food Prot. 59 (10): 1127-1132.
[82] WHO Health Education in Food Safety. World Health Organization, Geneva (in press).
[83] WHO (1981) Wholesomeness of Irradiated Food: Report of a Joint FAO/IAEA/WHO Expert Committee, World Health Organization Technical Report Series No. 659, Geneva.
[84] WHO (1984) The Role of Food Safety in Health and Development, World Health Organization Technical Report Series No. 705, Geneva.
[85] WHO (1994) Safety and Nutritional Adequacy of Irradiated Food, World Health Organization, Geneva.
[86] WHO (1995a) Application of Risk Analysis to Food Standard Issues, Report of a Joint FAO/WHO Expert Consultation, World Health Organization, Geneva.
[87] WHO (1995b) Food Technologies and Health, Food Safety Issues WHO/FNU/FOS/95.12, World Health Organization, Geneva.
[88] WHO (1997a) World Health Report 1997, World Health Organization, Geneva.
[89] WHO (1997b) Food Irradiation - Sky´s the Limit, Press Release WHO/68, 19 September, 1997, Geneva.

TABLE 1 RADIATION D_{10} VALUES FOR VARIOUS PATHOGENIC NON-SPORE FORMING BACTERIA IN FOODS

Microorganism	Food	D_{10} (Kgy)
Campylobacter jejuni	Beef filet	0.08-0.11
	Ground beef	0.16-0.32
Vibrio cholerae	Prawns	0.11
Yersinia enterocolitica	Ground beef	0.10-0.21
	Minced pork	0.16-0.19
Aeromonas hydrophila	Shrimp paste	0.21
	Ground beef	0.14-0.19
Escherichia coli	Minced chicken	0.26-0.39
Escherichia coli O157:H7	MD chicken	0.26-0.44
	Ground beef	0.24-0.30
Vibrio vulnificus	Shrimp paste	0.30
Salmonella typhimurium	Beef filet	0.37
	Ground beef	0.67
	Roast beef	0.44-0.55
Staphylococcus aureus	Roast beef	0.39
	Ground beef	0.43-0.45
Vibrio parahaemolyticus	Shrimp paste	0.44
Listeria monocytogenes	Minced pork	0.57-0.70
	MD Chicken	0.27-0.77
	Ground bccf	0.50-0.61

Source: Thayer (1995); Thayer "et al.," (1995); Loaharanu (1996).

TABLE 2 MINIMUM EFFECTIVE DOSE FOR RENDERING SOME FOODBORNE PARASITES UNINFECTIVE

Organism	Minimum Effective Dose (Kgy)
Toxoplasma gondii	0.5
Fasciola hepatica	0.7
Clonorchis spp.	0.15
Angiostrongylus cantonensis	2.0
Cysticercus bovis (Taenia saginata)	0.4
Cysticercus cellulosae (Taenia solium)	0.2-0.6
Trichinella spiralis	0.1-0.3

Source: IAEA (1993); Roberts and Murrell (1993); Loaharau (1996).

Role of Irradiation in a Multiple-Hurdle Approach to Food Safety

M. PATTERSON
Food Science Division
Department of Agriculture for Northern Ireland and
The Queen's University of Belfast
Newforge Lane
Belfast BT9 5PX, Northern Ireland, UK

ABSTRACT

Irradiation is a very effective preservation method that can improve the microbiological safety and quality of many foods. However, the dose which can be applied, and hence the extent of microbial kill, may be limited by undesirable changes in sensory quality including flavour, appearance and texture. This problem can be overcome by using irradiation in combination with other preservation techniques. The idea of superimposing a number of preservation methods has been described as the "Hurdle Concept". It was originally used for the formulation of a number of shelf-stable meat products but the concept can be applied to a wide range of foods, particularly as minimally processed, low-additive products are perceived by consumers as being more "natural" and "healthy". There are many examples of how the hurdle concept is used by the food industry to extend shelf-life and/or increase microbiological safety of foods as well as making them shelf-stable. This paper will describe a number of such applications involving the use of irradiation in combination with other preservation methods to enhance preservative action and/or reduce the severity of one or all the treatments. Some would argue that this approach is where the major benefits of food irradiation will eventually be seen.

One of the most common combination treatments is the use of irradiation in with refrigeration. For example, a dose of 2.5-3 kGy is sufficient to significantly reduce vegetative pathogens in poultry meat. Many spoilage organisms will also be killed, thus extending the shelf-life. However, the maximum potential of shelf-life extension will only be achieved if the meat is kept refrigerated after irradiation. This same principle also applies to milk pasteurisation, where the heat treatment is primarily aimed at inactivating vegetative pathogens and, as numbers of spoilage organisms are also reduced, the product has an increased shelf-life. However, in order to achieve most benefit from the process the additional hurdle of refrigeration has to be used.

A combination of irradiation with mild heat has also been shown to be effective in controlling spoilage organisms and increasing the safety of certain foods. For example, an irradiation dose of 5 kGy is sufficient to sensitise clostridial spores to subsequent heating . This was found to be of value in heat-treated spiced meats, where the heat treatment required in the canning process could be reduced as the spores in the irradiated spices were heat sensitised. Irradiation can also sensitise vegetative cells to subsequent heating. This could be advantageous in irradiated foods, such as raw meats, which are cooked before consumption and in ready meals, which are reheated prior to consumption. Other physical treatments, such as modified atmosphere packaging can also act synergistically with irradiation to improve microbial kill.

Successful combinations of chemical preservatives with irradiation have been reported by several authors. For example, the addition of sodium chloride or nitrite along with irradiation was found to delay or inhibit toxin formation by *C. botulinum.*

In conclusion, the preservative effects of radiation can often be advantageously combined with the effects of other physical or chemical agents. The resulting multi-hurdle treatments may involve synergistic or cumulative actions, leading to a decreased requirement for one or more of the agents. This in turn may result in improved sensory and microbiological quality of the products.

Introduction.

There are a wide variety of preservation methods available for use by the food industry. Traditionally the main aim of these methods has been to control the growth of pathogenic and spoilage microorganisms. Increasingly, however, consumers are demanding other quality attributes in addition to foods that are microbiologically safe and with an extended shelf-life. Consumer expectations of quality varies but can include the concepts of good sensory quality (flavour, texture, aroma, appearance), good nutritional quality (for example, retention of vitamins), a natural, fresh appearance and absence of chemical additives. It is impossible to find a single preservation method that can meet all these demands in all foods. It is therefore necessary to consider each food on an individual basis and choose the most appropriate preservation method to achieve the desired purpose.

Only a few preservation techniques act by destroying all the microorganisms present in foods e.g. heat and irradiation sterilization. Such methods usually require an extreme use of the treatment and this may result in a loss of product quality. Alternatively, a less extreme use of these technologies, such as heat pasteurisation or low dose irradiation can kill some microorganisms, especially vegetative pathogens and spoilage bacteria but the products will have a relatively short shelf life and may still result in a loss of quality in some products. Other methods, such as modified atmosphere packaging (MAP), vacuum packaging, acidification or addition of chemicals, act by inhibiting the activity of microorganisms to varying degrees but not actually killing them outright.

It has been found that the use of a number of these preservation methods in combination can be advantageous, especially by allowing a less extreme use of any single treatment, leading to improvement in product quality (1). This approach of superimposing a number of limiting factors was described as the "Hurdle Concept" by Leistner (2). He regarded each preservation factor as a hurdle but in reality the situation is more complex as many factors interact synergistically rather than additively.

Leistner originally used the concept in the formulation of a number of shelf-stable meat products but it is widely used in the food industry. For example, nitrite with mild heat treatment are used to control the growth of, and toxin production by, *Clostridium botulinum* in canned meats. Sugars such as sucrose are used with a heat treatment in the making of fruit preserves and condensed milk. The hurdle concept is particularly relevant in recent years because additive free, minimally processed products are perceived by consumers to be more "natural" and "healthy".

This paper will present examples of applications of the multiple-hurdle approach involving the irradiation of a variety of foods for quality and safety. Some of the applications, such as irradiation with chilled storage are already widely used. Other applications, such as irradiation with modified atmosphere packaging, show good potential and promise for the near future. Further details of other examples of combination treatments involving irradiation can be found in a recent IAEA publication (3).

Combination treatments to improve the safety and quality of meat, fish and poultry.

2.1 Treatments involving chill temperatures

Irradiation is known to be successful in improving the microbiological quality of meat, fish and poultry. Many commercial applications rely on low dose irradiation combined with chilled storage during and after treatment. The outcome, in terms of microbiological quality, is similar to that achieved by heat pasteurisation. The low dose irradiation is sufficient to kill vegetative bacterial pathogens and many spoilage microorganisms. However, the product is not sterile and requires to be stored under chilled conditions to achieve the optimal effect. For example, an irradiation treatment of 2.5 kGy is sufficient to cause at least a 100,000 fold reduction in numbers of pathogens such as *Salmonella* and *Campylobacter* as well as spoilage organisms such as *Pseudomonas* spp. The shelf-life of the meat will be at least 14 days if stored at 3°C but less than 8 days if stored at 10°C. The US Food and Drug Administration recently approved the irradiation of chilled red meat to a maximum dose of 4.5 kGy and frozen meat to a maximum dose of 7 kGy (4). The main purpose of the treatment is to significantly reduce the numbers of food borne pathogens, such as *Escherichia coli* O157:H7 with the secondary benefit of extending shelf-life, providing good temperature control is maintained.

2.2 Treatments involving heating

The use of irradiation in combination with heat was first suggested in the 1950s when synergistic effects were observed for the killing of bacteria. More recently a number of applications have been suggested to control spoilage and increase the microbiological safety of certain foods (5). There is still some debate about the order of application of the heat and irradiation treatments and whether they should be applied separately or simultaneously. However, it is generally agreed that the combined treatments have a synergistic effect on vegetative bacteria and bacterial spores. It has been shown that even if sub-lethal doses of irradiation are applied, the surviving microorganisms can be injured and are therefore more susceptible, or sensitised, to other preservation treatments. Thayer *et al* (6) reported that low-dose radiation sensitised *Salmonella typhimurium* in mechanically deboned poultry meat to subsequent heat treatment. The effect persisted for up to 6 weeks. This synergy between heat and irradiation, coupled with the persistence of the radiation-induced heat sensitisation could be used to particular advantage in irradiated foods which have to be reheated prior to consumption, such as ready meals. A low dose irradiation treatment during processing would result in a greater than expected kill of any surviving microorganisms during the reheating stage. This hypothesis was tested in a study where samples of minced cook-chill roast beef and gravy were inoculated with 10^9 cells/g of *Listeria monocytogenes*, *S. typhimurium* or *S. seftenberg* and then subjected to heating at 60°, 65° or 70^oC, or heating after irradiation to a dose of 0.8 kGy (7). The number of surviving colony forming units were enumerated by plating on a non-selective recovery medium containing pyruvate to aid the recovery of sublethally injured cells. In the case of the *L. monocytogenes* strains, irradiation prior to heating caused a significant reduction in the Decimal reduction (D_{10}) value indicating that the cells were more heat sensitive. Z values (^oC) were also calculated for the resulting thermal death time curves (log D against heating temperature). This provides information on the relative resistance of the organism to different temperatures and allows for the calculation of equivalent thermal processes at different temperatures. The Z values were lower in radiation-sensitised listeria cells indicating a decreased heat resistance. Subsequent studies have shown that the irradiation sensitising effect persists for at least 5 days between irradiation and heating. The irradiation sensitising effect was less pronounced for *S. typhimurium* and in the case of *S. seftenberg* irradiating prior to heating seemed to increase heat resistance. The results of this work have produced quantitative evidence of radiation-induced heat sensitisation in *L. monocytogenes* and *S. typhimurium*. The combined use of these two preservation treatments could improve the microbiological quality of certain foods.

Irradiation can also sensitise spores to subsequent heating and it has been proposed that this combination could be of value in heat-treated spiced meats (8). Spices often contain large numbers of bacterial endospores and doses greater than 10 kGy would be required for an acceptable reduction in numbers. However, doses below 5 kGy would be sufficient to sensitise the spores to heat and improve the microbiological quality. This would result in a substantial reduction in the heat treatment required for seasoned food. It was calculated that the heat process required to produce commercial sterility in canned luncheon meat could be reduced from an F_0 value of 4.7 to 3.4 if irradiated spices were used. This reduction in heating gave better product quality and a saving in energy (9).

2.3 Treatments involving modified atmospheres

The combination of irradiation with modified atmosphere packaging (MAP) to improve the microbiological quality of meat and fish products has been investigated by a number of research groups (10, 11, 12). Modified atmosphere packaging is already used in conjunction with chill storage to extend the shelf-life of some fresh and processed meats. Although MAP can slow down the growth of some undesirable microorganisms, including spoilage and pathogenic bacteria, it does not generally kill them. When the atmosphere is removed, for example when the pack is opened, microbial growth is resumed. The rationale for using a combination of low dose irradiation is as follows: irradiation can reduce the numbers of spoilage and pathogenic microorganisms and injure any survivors while the modified atmosphere will suppress the growth of the survivors during subsequent storage. Irradiating under a modified atmosphere can also act synergistically in the killing of bacteria. Some bacteria in poultry meat, such as *Moraxella phenylpruvica* and *Lactobacillus* sp. were more sensitive to irradiation when the treatment was carried out under vacuum or in 100% carbon dioxide (13). Similar results for lactobacilli irradiated in beef (14).

This combined approach can be particularly useful when applied to foods that are sensitive to irradiation off-odours and flavours, even at relatively low doses. Pork has a relatively low threshold dose of 1.75 kGy before undesirable organoleptic changes can be detected (15). A study of pork chops showed that a combination of irradiating to 1.75 kGy in packs containing 25% carbon dioxide and 75% nitrogen, followed by storage at 4°C gave acceptable products (in terms of colour and odour) for at least 17 days. This was compared to 8 days for chops packed in the same atmosphere but without irradiation (16). Inoculation studies using a variety of vegetative pathogens indicated that the microbiological safety of the irradiated MAP pork was better than that of the non-irradiated samples (17).

The microbiological safety of MAP products in general has been the subject of much discussion. The debate has extended to include irradiated MAP products. A particular concern has been that low-dose irradiation combined with MAP may inhibit many spoilage bacteria and so allow the growth of, and toxin production by, *Cl. botulinum* especially if there is poor temperature control while the product is still organoleptically acceptable (18). A further concern is the possibility a combination of irradiation with MAP may activate spore germination and enhance toxin production by *Cl. botulinum* if the product is temperature abused (19). These issues were addressed in a series of inoculation studies on pork, using varying irradiation doses (0, 0.5 or 1.0 kGy), modified atmospheres (0, 10 or 20% O_2, balance N_2) and storage temperatures (5, 15 or 25°C) (20). Storage temperature had the greatest effect on toxin production by *Cl. botulinum.* No toxin was detected in any sample stored at 5°C, emphasising the importance of proper refrigerated storage to prevent the growth of proteolytic strains of *Cl. botulinum.* All samples were toxic after 2 days at 25°C, irrespective of atmosphere or irradiation treatment. When storage was at 15°C, irradiation delayed toxin production in samples packaged under anaerobic conditions but when O_2 was present initially irradiation had no effect on toxin production.

Non-proteolytic strains of *Cl. botulinum* could have particular significance in products packaged under modified atmospheres or in a vacuum, especially if the numbers of competing microorganisms were reduced by irradiation. These strains are capable of growing and producing toxin at refrigeration temperatures without causing obvious spoilage. Firstenberg-Eden *et al* (21,22) inoculated chicken skins with *Cl. botulinum* type E spores before vacuum packaging and irradiating to 3 kGy. The samples were stored at 10 and 30°C. At both temperatures spoilage occurred before toxin was produced indicating that sufficient microorganisms survived the treatment to compete effective against the pathogen. However, when the experiment was repeated with an irradiation dose of 5 kGy, toxin formation occurred before spoilage in samples stored at 30°C. These results suggest that safety concerns over irradiated and vacuum packaged or MAP meat may only be valid when doses above 3 kGy are used.

There are several reports of the successful use of irradiation with MAP or vacuum packaging to improve the shelf-life of fish such as hake and cod (12). The problem of rancidity development in fatty fish such as petrale sole after irradiation can be overcome by excluding O_2 during irradiation and storage. However, the potential for the growth of and toxin production by *Cl. botulinum* type E is again of concern. Scientific evidence is conflicting and seems to depend on a number of factors including the chemical composition of the fish, storage temperature, initial spore numbers and methods used to estimate shelf-life (23). The irradiation dose is also important. Doses around 1 kGy have been proposed (24) so that sufficient spoilage microorganisms survive to compete with *Cl. botulinum* during storage. A storage temperature no greater than 3.3°C is required to prevent toxin production irrespective of packaging atmosphere.

In many studies of the effect of irradiation combined with vacuum packaging or MAP it has been noted that lactic acid bacteria form a significant part of the microbial population of the meat products during storage (16). Many lactic acid bacteria are known to be antagonistic towards pathogens. Thus, their growth in irradiated MAP and vacuum packs during storage could confer additional safety to these products.

2.4 Treatments involving chemical preservatives

Various studies have investigated the combination of irradiation and chemical treatments. For example, combining irradiation (2.5 kGy) with dipping in acetic acid extended the shelf-life (based on microbiological and sensory quality) of buffalo meat by up to 4 weeks compared to non-irradiated controls (25). A combination of irradiation (2 kGy), pH reduction to 5.6 and water activity to 0.96 extended the shelf-life of vacuum packaged ready to fry chilled meat products by a factor of 3 (26). Lactic acid bacteria were the main component of the surviving microflora in these products. Gamma radiation at 2 kGy in combination with reduced water activity and pH resulted in a 2 decimal reduction in the aerobic cell count and a 4 decimal reduction in Enterobacteriaceae in a chilled meat product (26). *Nephrops norvegicus* (cold water prawn) normally has a short shelf-life due to microbial spoilage and the development of melanosis (black spot) from the action of polyphenol oxidase. Irradiation will not inactivate the enzyme but activity can be inhibited by 4-hexylresorcinol, a compound with GRAS approval. A combination of 0.005% dip of 4-hexylresorcinol for 1 minute followed by irradiation (1.6 kGy) was sufficient to inhibit melanosis and prevent microbial spoilage of whole *Nephrops* for at least 15 days at 3°C.

Combination treatments to improve the safety and quality of fruit and vegetables.

Combination treatments involving the irradiation of fruit and vegetables tend to be aimed at improving quality rather than safety. For example, combination treatments using hot water dipping followed by low dose irradiation have been proposed to improve the quality of certain products. Irradiation can inhibit sprouting of white potatoes at doses of 0.1 kGy but spoilage due to microbial rot can occur at high ambient temperatures (25-30°C) which are common in tropical countries. A reduction in microbial spoilage can be achieved by first heating the potatoes in water for 10 minutes at 85°C (27). Similarly, a combination of hot water dipping (55°C for 20 minutes) followed by an irradiation dose of 0.6 kGy extended the shelf-life of mangoes stored at 18-20°C from 15 to 32 days. The sensory quality (as assessed by a consumer panel) was not acceptable and the irradiated mangoes were actually found to be sweeter and more tender. The levels of vitamins A and C were unaffected by the combination treatment (28).

Combination treatments to improve the safety and quality of chilled and shelf-stable meals.

Ready-prepared meals are becoming increasingly popular. The quality and variety of such meals has improved over the years and, as they are already cooked and only require reheating, they are seen as being very convenient in the busy lives of consumers. Traditionally ready-prepared meals are frozen to maintain shelf-life. However, freezing is a relatively expensive technology. Also, consumers perceive chilled products to be more acceptable than frozen. Chilled ready meals have not only become more popular in the general retail sector but also in mass catering, for example for hospitals, day care centres and residential homes for the elderly. In many cases large numbers of the meals are prepared in a central facility and then transported to various locations. The meals normally have a relatively short shelf-life (5 days including day of production and consumption) and could potentially become contaminated with pathogens during production. Irradiation has the potential to improve the microbiological safety and quality of these foods. Pathogens can be destroyed and numbers of spoilage organisms can be reduced, potentially giving longer shelf-life and so reducing the need for daily deliveries to outlets (residential homes, day care centres etc).

In one study gamma radiation of ready-to-eat chilled meals (roast beef and gravy, roast potatoes, carrots and broccoli) was successful in extending shelf-life and improving microbiological safety without significantly affecting sensory properties (29). Thus, the combined effects of cooking, irradiation and chilled storage worked successfully to give high quality, safe products. Other studies have investigated the use of much higher irradiation doses, sufficient to sterilize foods. Irradiation sterilized meals have been consumed by astronauts for over 20 years because they are perceived to be of a higher quality than foods produced by other preservation methods. However, other sections of the population increasingly need sterile foods. For example, people who are immunocompromised (HIV positive, cancer and transplant patients) or who are elderly are regarded as being at higher risk from infection from contaminated foods. Also, niche markets are developing, such as the need for sterile foods for use by disaster victims and campers.

A Study Group convened in 1997 by the Joint Food and Agriculture Organisation of the United Nations (FAO) / International Atomic Energy Agency (IAEA) / World Health Organisation (WHO) discussed the safety and nutritional consequences of using high doses of irradiation to treat foods. The group concluded that high dose irradiation could be used safely to produce sterile foods. No upper limit of dose was recommended. Gamma radiation is used in South Africa to produce shelf-stable ready meals. Here an overall average dose of 45 kGy is used and the meals are cooked and frozen before irradiation. This combination minimises adverse sensory changes which can occur when high doses of irradiation are used (see I. De Bruyn in this volume).

It has also been demonstrated that electron beam treatment at doses ranging from 12 to 40 kGy combined with cooking, vacuum packaging and freezing (-40°C) results in meals remaining microbiological acceptable for up to 11 months of storage at 4°C. The meals included duck in orange sauce, chicken, salmon and white fish, fried potatoes, pancakes, rice, mashed potatoes and carrots. All the meals were regarded as being organoleptically acceptable, except the fried potatoes which became soft and developed an off-odour during storage (30).

Conclusions

There is much evidence to show that irradiation in combination with other preservation treatments can be used very successfully to improve the quality and safety of many foods. However, as with all food processing operations, the conditions must be optimised for each product. In addition, it must be remembered that no preservation method, including irradiation, should be substituted for Good Manufacturing Practice.

References

1. Banks, J.G. 1987. "The perfect combination". Food, January:36-39.
2. Leistner, L. 1985. "Hurdle technology applied to meat products of the shelf-stable product and intermediate moisture food types". Properties of water in Foods eds D. Simatos and J.L. Multon. Dordrecht: Martinus Nijhoff Publishers, pp 309-329.
3. Anon. 1998. Combination Processes for Food Irradiation. Vienna: International Atomic Energy Agency.
4. Food and Drug Administration. 1997. "Irradiation in the production, processing and handling of food". Federal Register, 62(232):64107-64121.
5. Farks, J. 1990. "Combination of irradiation with mild heat treatment". Food Control, 1:223-229.
6. Thayer, D.W., Songprasertchai, S. and Boyd, G. 1991. "Effects of heat and ionizing radiation on *Salmonella typhimurium* in mechanically deboned chicken meat". J. Food Protection, 42:410-415.
7. Grant, I.R. and Patterson, M.F. 1994. "Combined effect of gamma radiation and heating on the destruction of *Listeria monocytogenes* and *Salmonella typhimurium* in cook-chill roast beef and gravy". Int. J. Food Microbiology, 27:117-128.
8. Kiss, I. and Farkas, J. 1981. "Combined effect of gamma irradiation and heat treatment on microflora of spices". In: Combination Processes in Food Irradiation. Vienna. International Atomic Energy Agency, pp 107-115.
9. Kiss, I., Zachariev, G.Y., Farkas, J., Szabad, J. and Tothpesti, K. 1978. "The use of irradiated ingredients in food processing". In: Food Preservation by Irradiation. Vol 1. Vienna. International Atomic Energy Agency, pp263-272.
10. Niemand, J.G., Van Der Linde, H.J. and Holzapfel, W.H. 1981. "Radurization of prime beef cuts". J. Food Protection, 44:677-681.
11. Ehioba, R.M., Kraft, A.A., Molins, R.A., Walker, H.W., Olson, D.G., Subbaraman, G. and Skowronski, R.P. 1987. "Effect of low dose (100 krad) gamma radiation on the microflora of vacuum-packaged ground pork with and without added sodium phosphates". J. Food Science, 52:1477-1480, 1505.

12. Lescano, G., Kairiyama, E., Narvaiz, P. and Kaupert, N. 1990. "Studies on quality of radurized (refrigerated) and non-radurized (frozen) hake (*Merluccius merluccius hubbsi*)". Lebensmittel-Wissenshaft und Technologie, 23:317-321.
13. Patterson, M.F. 1988. "Sensitivity of bacteria to irradiation on poultry meat under various atmospheres". Letters Applied Microbiology, 7:55-58.
14. Hastings, J.W., Holzapfel, W.H. and Niemand, J.G. 1986. "Radiation resistance of lactobacilli isolated from radurized meat relative to growth and environment". App. Environmental Microbiology, 52:898-901.
15. Sudarmadji, S. and Urbain, W.M. 1972. "Flavour sensitivity of selected animal protein foods to gamma radiation". J. Food Science, 37:371-372.
16. Grant, I.R. and Patterson, M.F.1991. "Effect of irradiation and modified atmosphere packaging on the microbiological and sensory quality of pork stored at refrigeration temperatures". Int. J. Food Science and Technology, 26:507-519.
17. Grant, I.R. and Patterson, M.F. 1991. "Effect of irradiation and modified atmosphere packaging on the microbiological safety of minced pork stored under abuse conditions". Int. J. Food Science and Technology, 26:521-533.
18. Lambert, A.D., Smith, J.P. and Dodds, K.L. 1991 "Shelf-life extension and microbiological safety of fresh meat – A review". Food Microbiology, 8:267-297.
19. Lebepe, S., Molins, R.A., Charoen, S.P., Farrar IV, H. and Skowronski, R.P. 1990. "Changes in the microflora and other characteristics of vacuum packaged pork loins irradiated at 3 kGy". J. Food Science, 55:918-924.
20. Lambert, A.D., Smith, J.P. and Dodds, K.L. 1991. "Combined effect of modified atmosphere packaging and low dose irradiation on toxin production by *Clostridium botulinum* in fresh pork". J. Food Protection, 54:94-101.
21. Firstenberg-Eden, R., Rowley, D.B. and Shattuck, G.E. 1982. "Factors affecting growth and toxin production by *Clostridium botulinum* type E on irradiated (0.3 Mrad) chicken skins". J. Food Science, 47:867-870.
22. Firstenberg-Eden, R., Rowley, D.B. and Shattuck, G.E. 1983. "Competitive growth of chicken skin microflora and *Clostridium botulinum* type E after an irradiation dose of 0.3 Mrad". J. Food Protection, 46:12-15.
23. Reddy, N.R., Armstrong, D.J., Rhodehamel, E.J. and Kautter, D.A. 1992. "Shelf-life extension and safety concerns about fresh fishery products packaged under modified atmospheres: A review". J. Food Safety, 12:87-118.
24. Urbain, W.M., ed. 1986. "Marine and Freshwater animal foods" In: Food Irradiation. London, Academic Press, pp 145-169.
25. Paul, P., Chawla, S.P. and Kanatt, S.R. 1998. "Combination of irradiation with other treatments to improve the shelf-life and quality of meat and meat products". In: Combination Processes for Food Irradiation. Vienna. International Atomic Energy Agency, pp111-130.
26. Farkas, J., Adrassy, E., Nanati, D., Horti, K., Mesaroa, L. and Reichart, O. 1998. "Interaction of ionizing radiation with other physical and chemical factors in improving the safety and storage stability of foods". In: Combination Processes for Food Irradiation. Vienna. International Atomic Energy Agency, pp15-32.
27. Urbain, W.M., ed. 1986. "Fruits, vegetables and nuts". In: Food Irradiation. London, Academic Press, pp 170-216.

28. Noomhorm, A., Ilangantileke, S.G., Upadhyay, I.P., Karki, D.B. and Apintanapong, M. 1998. "Use of irradiation in combination with preservation techniques to extend the shelf-life of tropical fruits ant their products". In: Combination Processes for Food Irradiation. Vienna. International Atomic Energy Agency, pp 53-76.
29. Patterson, M.F., Stevenson, M.H., Grant, I.R., McAteer, N.J. and Stewart, E.M. 1998. "Effect of gamma irradiation on the microbiological and organoleptic quality of ready prepared chilled meals". In: Combination Processes for Food Irradiation. Vienna. International Atomic Energy Agency, pp 159-180.
30. Desmonts, M-H., Pierrart, N., Ingersheim, A., Strasser, A. and Hasselmann, C. 1998. "Use of irradiation in combination with other treatments to improve the quality of precooked meals". In: Combination Processes for Food Irradiation. Vienna. International Atomic Energy Agency, pp. 201-222.

Reducing Microbial Contamination Including Some Pathogens in Minced Beef by Irradiation

I. F. KISS and L. MÉSZÁROS
University of Horticulture and Food Industry
Budapest, Hungary

H. KOVÁCS-DOMJÁN
National Food Investigation Institute
Budapest, Hungary

Abstract

The aim of our experiments was to reduce some pathogenic and spoilage microorganisms in aerobically packed minced beef by ionizing radiation. The microbial contamination of beef is relatively low. Application of the Good Manufacturing and Hygienic Practice improves the quality and safety of foods but the storability of minced meat at 4°C is very short. The growth of microorganisms including that of the very few pathogens too, is quite fast, therefore this type of meat product may represent potential health hazard. Ionizing radiation is one of the possible treatments for decontamination of meat, it is a physical method. Radiation can inactivate microorganisms in foods without significant alteration of their sensorial acceptability. We studied the effect of irradiation in aerobically-packed minced beef with natural and high contamination (10^4 and 10^8 g^{-1} initial cell count) and inoculated with a cocktail suspension of *E.coli, S.aureus and L.monocytogenes* (each strain 10^6-10^7 CFU cm^{-3}). Polyethylene foil packed samples were irradiated with 1, 2 and 3 kGy and stored at 4°C for 10 days. The survival curves and the D_{10} values of inoculum-strains and natural microflora, as well as the shelf-life values were determined. We concluded from the data that a dose of 2 kGy extends the shelf-life 3-5-times and the inoculated pathogenic strains can be reduced by 7, 5 and 5 log cycles, respectively. During 10 days of chilled storage at 4°C, the reduced viable cell counts did not change except that of *L.monocytogenes*, which started to grow after 3 days.

INTRODUCTION

The safety and keeping quality of foods are very important requirements to the food industry. Therefore, food technologies have a very important role to preserve foods, reduce losses and ensure the safety of products. Foods are produced by the agriculture area in natural environments, and in the food industry, where microorganisms are very common. These microorganisms are contaminating food products very often during their production, processing and handling. One part of the microbiota consists of spoilage causing microbes and the other part may be pathogenic. The microflora is mostly very characteristic for different food groups taking into consideration the type of spoilage, just as the symptoms of illnesses caused by pathogenic microorganisms. An important requirement is for the food industry that the product shall not contain harmful microorganisms or only at such a low level which does not create an unacceptable health risk. The Good Manufacturing and Good Hygienic Practices are able to reduce the losses and risk, and prevent microbiological contamination which causes the largest part of spoilage and is a very important reason of foodborne illness, because foods are transmitting pathogenic microorganisms.

Some food technologies inhibit the activity of microorganisms whereas others reduce viable cell numbers of them by killing, but prevention of the contamination of foods is not always possible. Therefore, the decontamination of foods is sometimes necessary. The microbiological contamination can be dangerous to health, pathogenic microorganisms are causing serious illnesses and the growth of spoilage microorganisms reduces the shelf-life and causes high economic losses. It is a well known fact that raw foods of animal origin are frequently contaminated with pathogenic microorganisms and in general less safe than vegetable foods.

During the last decades the number of diseases of microbiological origin transmitted by foods was increasing from year to year. Microorganisms are submitted through foods and water and, of course, from one human to another. The lifestyle is also changing as well as the size of food production and distribution systems. Among pathogenic microorganisms the priority has given worldwide to problems related to *Salmonella, Campylobacter, S.aureus, C.perfringens, C.botulinum, B.cereus, Shigella, Vibrio, E.coli* and more and more trouble is caused by *Listeria monocytogenes* (20, 18, 16). The microbial contamination is not limited to developing countries but it is present in developed countries, too. These facts stimulate scientists, food industry and health authorities to improve the microbiological quality of foods and to apply very effective methods to reduce the number of microorganisms of foods. The microbiological decontamination is one step in a process, i.e. processing for safety should be an integrated part of some technologies. This treatment can be a chemical, physical or combined method depending on properties of food, type of contamination, technology and efficacy of the applied decontamination method (1, 6, 8, 9, 10, 17, 19).

Ionizing radiation is one of the possible treatments, it is a physical method, effective without heat effect, as a terminal treatment. One of the most attractive decontamination methods is radicidation (14). Most of food pathogens are sensitive to irradiation and the number of spoilage causing microorganisms are reduced by relative low doses, therefore, the radicidation and radurization are destroying pathogens and extend shelf-life of meat (4, 14).

Meat and poultry are perishable foods which is partly connected with their chemical composition and very active microflora. Their shelf-life is relatively short, 3-5 days at refrigeration temperatures. The initial contamination ranges between 10^3 – 10^5 CFU g^{-1}, depending from slaughter-house technology, hygienic conditions, technological and hygienic practices. The shelf-life of minced meat is very short, 1 day. Microorganisms are well distributed in the meat during preparing the product, which may be contaminated with *Salmonella* and other pathogens. The radicidation can help in preventing of meat-borne diseases. Low-dose irradiation will reduce the level of microorganisms of concern to public health considerably, especially after pre-packing of the cuts (5). The aim of our experiments was to reduce some pathogenic and spoilage microorganisms in aerobically packed minced beef by ionizing radiation.

METHODS

Material

Minced beef was divided into two parts. One had the original total microflora (OTVC) and other part was inoculated with an overnight broth culture cocktail (containing 10^6-10^7 CFU per millilitre from each strains of *E.coli* OKI 33001, ATCC 9637; *S.aureus* ATCC 9144; *L.monocytogenes* ATCC 19111, NCTC 7973). The inoculated was marked as TVC and the minced beef after its inoculation/homogenisation was packed in polyethylene pouches (aerobically-packed).

Irradiation

The radiation doses applied were 1.0, 2.0, and 3.0 kGy, respectively. Irradiation was carried out in a RH-gamma-30 (Co-60) laboratory source (dose rate 1.99 kGy per hour). Samples were irradiated at the "Frederic Joliot-Curie" National Research Institute for Radiobiology and Radiohygiene, Budapest.

Microbiological examinations

Samples were stored at 4°C. The microbiological investigations of inoculated and uninoculated samples were carried out as a function of radiation dose and storage time. A 10 g sample was homogenised in 90 ml pepton-sodiumchloride solution by a Stomacher and decimal dilution series were prepared. The total viable cell count (the original microflora – OTVC - and that inoculated with the cocktail – TVC) of samples were determined on Nutrient Agar (Oxoid) at 30°C for 48 h; *E.coli* on VRB Agar (Oxoid) plates at 30°C for 24 h; *S.aureus* on Baird-Parker Agar (Oxoid) plates at 37°C for 48 h; *L.monocytogenes*: Palcam Agar with Supplement (MERCK) at 37°C for 48 h.

Evaluation

Shelf-lifes (*L*). of unirradiated and irradiated samples were determined from viable cell counts considering 7.0 log CFU g^{-1} as a spoilage level. The survival fraction of *E.coli, Listeria monocytogenes* and *Staphylococcus aureus* were examined as relative changes of their actual viable cell counts, $\log_{10}(N/N_0)$. The D values of strains were determined from regression equations of relationships between survival fraction and radiation dose.

RESULTS

Shelf-life extension of minced beef.

The shelf-life of foods depends from the microbiological contamination. The initial cell count what can be reduced to lower level by a decontamination technique practically determines the storability of foods. Irradiation as a decontamination method has a double effect, on one side it reduces the viable cell count and influences the growth rate of the survival fraction, thereby the shelf-life is extended, and on other side, it reduces pathogenic microorganisms, too. The shelf-life extension of minced meat by irradiation has been investigated by many researchers and the suggested dose range is 1-5 kGy. In our experiments we found out that the useful dose is between 2-3 kGy in case of aerobically packaging in chilled form. At higher doses off-flavour of the product could be detected (7.). We carried out our experiments with a normal initial cell count (case *A* - natural contamination 10^4 CFU g^{-1}) and with a higher level of 10^8 CFU g^{-1} (case *B*) too. In the case of experiment *A* the natural contamination (original total viable cell count–OTVC) of minced beef was $3x10^4$ CFU g^{-1} and decreased by 1.1, 2.2 and 3.0 log cycles at 1.0, 2.0 and 3.0 kGy, respectively. Irradiated samples reached the critical value of, 10^7 CFU g^1 (*L*) after 3-, 5- and 6 days. As compared with unirradiated sample, the shelf-life increased 3-, 5- and 6-times (Fig.1.). When samples were inoculated with a cocktail of bacteria (*E. coli, S. aureus, L. monocytogenes)* marked TVC (total viable cell count) the shelf-life didn't change too much as compared with naturally contaminated samples (Fig.2.). The pathogens could influence the growth of normal flora because they were present in higher number than original flora. Spoilage of meats in aerobic-packages follows a fairly normal course with a delay corresponding to diminution of the initial contamination(4).

In the other experiments (*B*) when we used more highly contaminated ($4x10^7$ CFU g^1) beef, the viable cell count of irradiated samples grew slower (Fig.3.). Inoculated samples showed very similar curves (Fig.4.) but the high number of pathogenic microbes produced an additive effect to reach earlier the critical level.

D_{10} values of microorganisms.

Survival curves of total viable cell count of natural microflora and inoculated strains in minced beef were studied as a function of irradiation dose. The estimated D_{10} values from regression equations were 0.88 kGy for the original microflora, 0.33 kGy for *E. coli,* 0.39 kGy for *S. aureus,* 0.43 kGy for *L. monocytogenes* (Fig.5.). The number of pathogens on meat is low under normal condition and most of them are sensitive to irradiation. The greatest contamination problems in meat and meat-products world-wide are *Salmonella, Campylobacter* and the above mentioned bacteria. The D_{10} values of different strains of these microorganisms are between 0.25-0.70 kGy. Consequently, taking into consideration these D values, relatively low doses can reduce their number or eliminate these microorganisms in minced meat and extend the shelf-life, too.

Survival of strains during storage time

Viable cells of the *Escherichia coli* strain in aerobically-packed minced beef stored at 4 °C in unirradiated samples increased almost 2 log cycles, over 10^9 CFU g^{-1} , in the first 3 days, when they reached the stationary phase (Fig.6.). The viable cell count of 1 and 2 kGy-irradiated samples was reduced by 3 and 6 log cycles. When they were treated with 3 kGy, we couldn't detect viable cell count. The viable cell count didn't practically change during the 10 days storage time.

The number of CFU of *S. aureus* was reduced very little in case of unirradiated and 1 kGy-dose treated samples during 10 days but samples irradiated with 2 and 3 kGy were stabile (Fig.7.). The colony count of *Listeria monocytogenes* in unirradiated samples increased only very little. In irradiated samples in case of low, normal cell count (10^4 g^{-1}) after 1-2 days they increased slowly. These changes were not found in samples which had a high contamination (Fig.8.). The pathogens could influence the growth of normal flora because they were present in higher number than the original flora.

CONCLUSIONS

The microbiological contamination is mostly the main reason for food spoilage and among the food-borne microorganisms sometimes pathogens are present. The Good Manufacturing and Good Hygienic Practices can prevent the microbiological contamination and reduce the risk to minimum, however, but mostly a decontamination process should be built into the food technology. One of them is the ionizing radiation which is able to reduce the microbiological contamination with high efficiency.

The aim of our experiments was to extend the shelf-life of the aerobically-packed minced beef and to study the survival of some pathogens after irradiation during storage time. It was concluded that the D_{10} value of the natural microbiological contamination of minced beef is 0.88 kGy. Thereby the viable cell number of 10^4 CFU g^{-1} initial contamination could be reduced by 2 log cycles and the shelf-life could be extended up to three-to five-times as an effect of 2 kGy (Fig.1.). In other experiments we determined that this dose did not influence the organoleptic properties (7, 8).

Taking into consideration that the foods are transmitting microorganisms and sometimes pathogens exist in the microflora, the food industry has the responsibility for the safety of foods. The microbiological quality of foods should be ensured and the health risk should be reduced to a minimum by an effective food technology. Food-transmitted infections and intoxications today are serious health problems in nearly all parts of the world. Pathogens spread first of all by foods of animal origin and they are causing foodborne diseases in very high number and death in some per cents too (5, 14, 15, 19). However these facts are underreported. As we know, only about 10 % of cases may be recorded. Our aim was in this part of experiments to study the survival of three pathogens: *Escherichia coli, Staphylococcus aureus* and *Listeria monocytogenes,* which were irradiated in minced beef. *E. coli* strains are a common part of the normal facultative anaerobic microflora of the intestinal tracts of humans and warm-blooded animals, but some groups of these strains are causing diseases (18, 19). We concluded from our data that our strain was very sensitive in minced beef to ionizing radiation and its viable cell count was reduced by 7 log cycles at 2 kGy and its survival fraction did not change during 10 days storage period. *S.aureus*, one of the most prevalent causes of gastroenteritis world-wide, is a very common microorganism and the number of incidences of staphylococcal food poisonings are very large (19, 15). On the basis of our experiments viable cell counts of *S. aureus* inoculated in minced beef has been reduced with 5 log cycles as an effect of 2 kGy. During the storage time the residual number of colony forming units did not change. *L. monocytogenes* is an ubiquitous microorganism, which has been increasingly recognized as an important foodborne pathogen over the last 15 years, being a bacterium responsible for opportunistic infections. High sensitivity of elderly people, young children, pregnant women and immundepressed persons towards *L. monocytogenes* is noted (16). In our experiments we estimated that the number of *L.monocytogenes* was reduced with 5 log cycles by 3 kGy, and the irradiation retarded the growth of survivors for 3 days, but after then- independently from the applied dose- they started to grow and reached the level of log CFU 7-8 values before spoilage was apparent. In an earlier experiment, we observed similar data (12). In another work it was concluded that shelf-life of a minimally processed ready-to-fry minced meat could be extended by a low dose irradiation (2 kGy) but 10^3-10^4 CFU · g^{-1} levels of *L. monocytogenes* could not be completely inactivated (3).

Finally, we are concluding that the ionizing radiation dose of 2 kGy extends the shelf life of aerobically-packed minced beef at 4°C storage temperature 3-5 times as compared with unirradiated samples. The same dose is reducing *E. coli* OKI 33001, ATCC 9637, *S. aureus* ATCC 9144 and *L. monocytogenes* ATCC 19111, NCTC 7973 strains 7, 5 and 5 log cycles, respectively in minced beef and the survivors of the first two strains didn't grow during 10 days storage time, but those of *L. monocytogenes* did after 3 days.

We thank Dr.L.Gazsó for the irradiation in "Frederic Joliot-Curie" National Research Institute for Radiobiology and Radiohygiene. The experiments were sponsored by the National Science Foundation of Hungary (OTKA No. T 021140).

REFERENCES

1. CHENG-AN, HWANG, BEUCHAT, L. (1995): Efficacy of selected chemicals for killing pathogenic and spoilage microorganisms on chicken skin. *J.Food Protect.* 58 (1):19– 23.
2. FARKAS , J (1997):.Physical methods of food preservation, in: Food Microbiology, Fundamentals and Frontiers (Eds.: Doyle, M.P., Beuchat, L.R., Monteville, T.J.) ASM Press, Washington D.C. pp.497– 515.
3. FARKAS, J., ANDRÁSSY, É. (1996): Behaviour of *Listeria monocytogenes* in an extended shelf-life chilled meat product. Acta Alim. 25 (2) :185-187.
4. INGRAM, M., FARKAS, J. (1977): Microbiology of foods by ionising radiation. *Acta Alim.*, 6 (2) :131.
5. KAMPELMACHER, E.H. (1981): Prospect of eliminating pathogens by the process of food irradiation. in: Combination Processes in Food Irradiation, Proc.Symp. Colombo, 24-28 November 1980. IAEA, Vienna, 1981. pp.265 – 289.
6. KISS, I., FARKAS, J. (1972): Radurization of whole eviscerated chicken carcass. *Acta Alim.* 1 (1):73 – 86.
7. KISS, I., ZACHARIEV, Gy. (1981): Use of irradiation to increase shelf–life of minced-meat product. in: Proc. 27. Europ. Fleischforsch. Kong. Wien, 1981. (Ed.:Prändl, O.) Wien, Vol. I.pp. 344-346.
8. KISS, I.F., BECZNER, J., ZACHARIEV, Gy., KOVÁCS, S. (1990): Irradiation of meat products, chicken and use of irradiated spices for sausages. *Radiat.Phys.Chem* .36 (3):295-299.

9. KISS, I.F., SVEIZER, Á., BECZNER, J., TÓTH, Á., KAFFKA, K., FÁBIÁN, Á. (1994) shelf-life extension of packed meat by irradiation. in: Food Irradiation in the Middle East and Europe, IAEA-TECDOC-754 , IAEA, Vienna, pp.59-75.
10. KISS, I.F., FARKAS, J., KISS-BRÜCKNER, M., SIMON, A., ANDRÁSSY, É., EL-AMAM, M. (1955): Effect of trisodium phosphate dip on survival and growth of spoilage flora and *Listeria monocytogenes* on skin of chilled broiler carcasses. in: Poultry Products Microbiology European Regulations and Quality Assurance Systems, Proc.XII.Europ.Symp.Quality of Eggs and Egg Products, Proc.2.Ann.Meet.EC COST WG No.2. 25-29 Sept.1995.Zaragoza, Spain, (Ed.:Briz,R.C.) iberCaja-COST-GOBIERNO De Aragon, pp.27-35.
11. KISS, I.F., MOHÁCSI-FARKAS, Cs., REICHART, O., MÉSZÁROS, L. (1998): Prediction of microbiological shelf-life of chilled irradiated meat. in: Shelf-life Prediction for Improved Safety and Quality of Foods, COPERNICUS Project CIPA-CT94-0120, COPERNICUS WS: "Shelf-life prediction of foods" 27-28 November 1997. Wageningen, pp.55-62.
12. KISS, I.F., FARKAS, J., KISS-BRÜCKNER, M., SIMON, A., ANDRÁSSY, É., Reduction of aerobic viable cell counts and *Listeria monocytogenes* inoculated onto chicken skin by ionizing radiation, in: Proc.New Technology for Safe and Shelf Stable Products, Sem. COST Action 97 Gr.3.(1998, Berlin), COST,pp.49 – 56.
13. KOVÁCS, S., KOVÁCS-DOMJÁN, H., KISS, I. (1981): Radiosensitivity of some micro-organisms on meat. in: Proc. 27. Europ. Fleischforsch. Kong. Wien, 1982. (Ed.:Prändl, O.) Wien, Vol.I. pp.347-350.
14. MOSSEL, D.A.A. (1985): Irradiation : An effective mode of processing food for safety. in: Food Irradiation Processing, Proc.Symp. Washington, D.C., 4 – 8 March 1985. IAEA, Vienna, 1985. pp.251 – 279.
15. RALOVICH, B. (1996): The problems of food-borne diseases in Hungary. *Acta Microbiol.Immun.Hung.* 43 (4):319 –332.
16. ROCOURT, J., BILLE, J. (1997): Foodborne listeriosis. *Wld hlth statist. quart.* 50(1/2):67 – 73.
17. SOMERS, E.B., SCHOENI, J.L., WONG AMY, C.L. (1994): Effect of trisodium phosphate on biofilm and planktonic cells *Campylobacter jejuni, Escherichia coli* O157:H7, *Listeria monocytogenes* and *Salmonella typhimurium. J.Food Microbiol.* 22 :269 – 276.
18. TAKEDA, Y. (1997): Enterohaemorhagic *Escherichia coli. Wld hlth statist. quart.* 50(1/2): 74 – 80.
19. TODD, E.C.D. (1997): Epidemology of foodborne diseases: a worldwide review. *Wld hlth statist. quart.* 50(1/2):30 – 50.

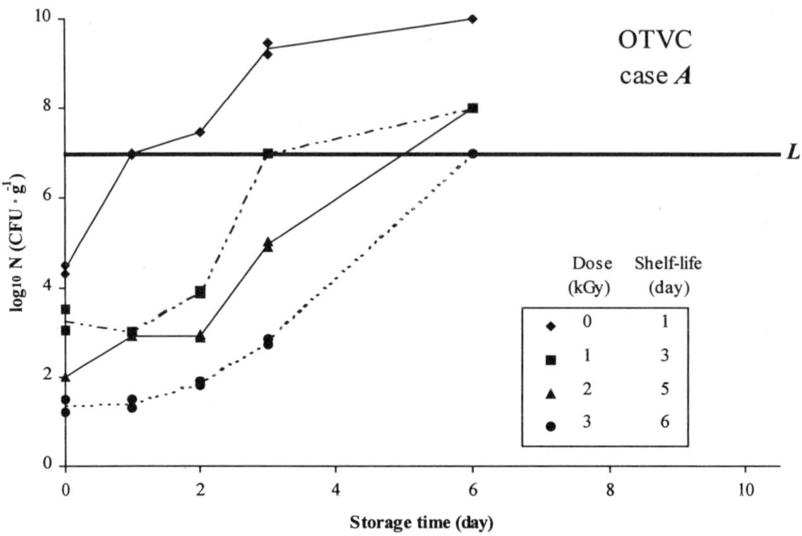

Fig. 1. Survival of microflora of aerobically packed minced beef with normal contamination (case *A*) as a function of irradiation dose and storage time at 4 °C. *L* - spoilage level.

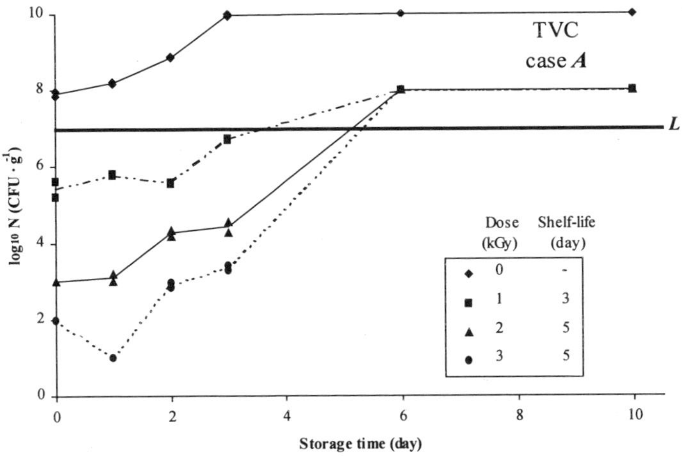

Fig. 2. Survival of microflora of aerobically-packed minced beef with normal contamination [inoculated with a bacteria cocktail (*E. coli, S. aureus, L. monocytogenes*)] (case *A*) as a function of irradiation dose and storage time at 4 °C. *L* - spoilage level.

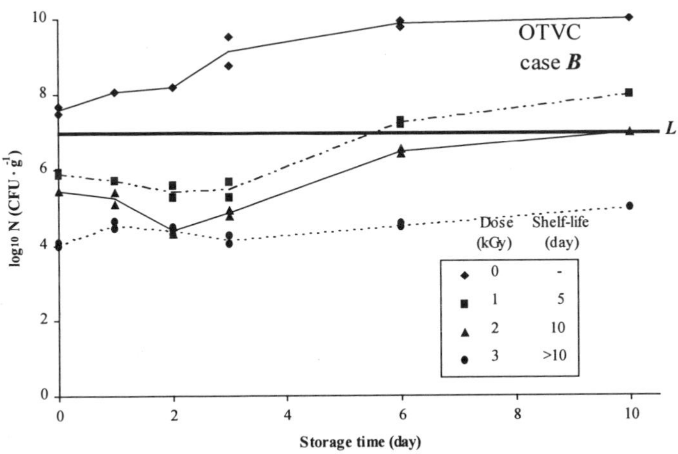

Fig. 3. Survival of microflora of aerobically packed minced beef with high contamination (case *B*) as a function of irradiation dose and storage time at 4 °C. *L*- spoilage level.

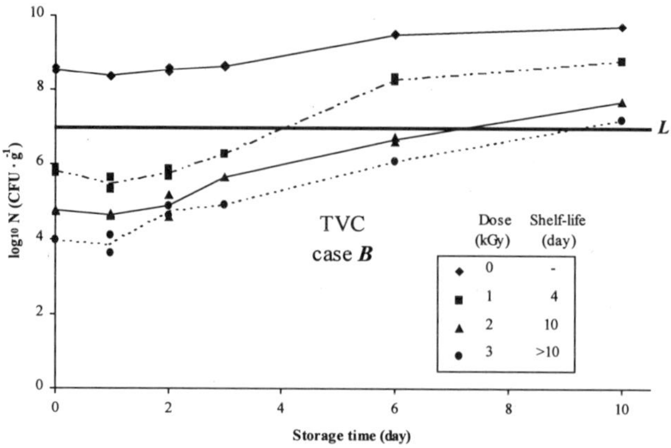

Fig. 4. Survival of microflora of aerobically-packed minced beef with high contamination [inoculated with bacteria cocktail (*E. coli, S. aureus, L. monocytogenes*)] (case *B*) as a function of irradiation dose and storage time at 4 °C. *L* - spoilage level.

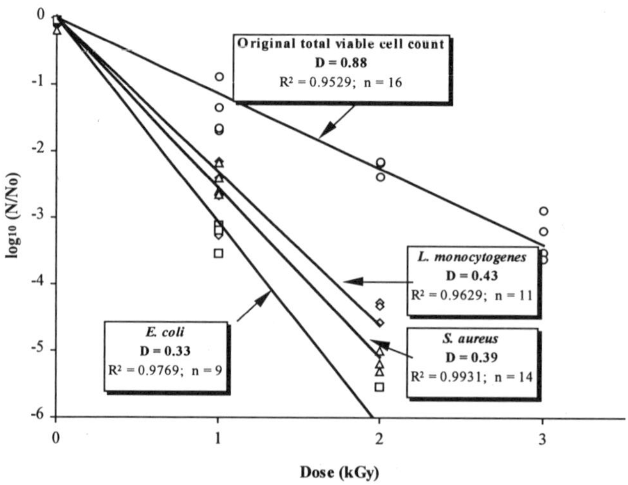

Fig. 5. Survival of original microflora and inoculated strains (*L. monocytogenes, S. aureus* and *E. coli*) in aerobically packed minced beef samples as a function of irradiation dose.

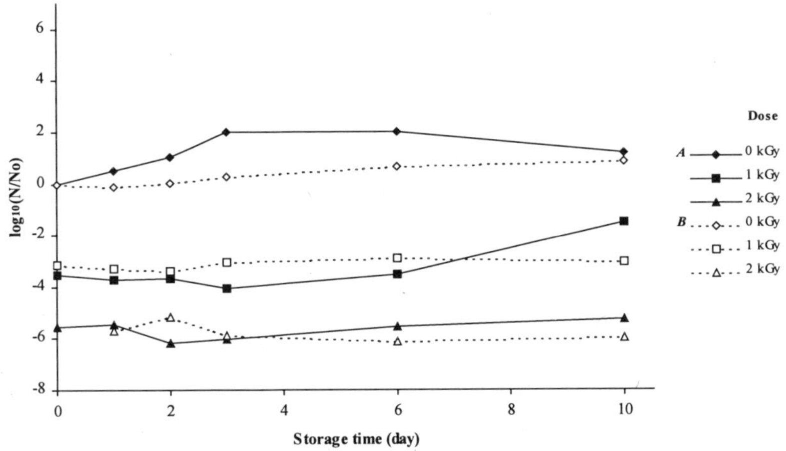

Fig. 6. Survival of *E. coli* in aerobically packed unirradiated and irradiated (1, 2 and 3 kGy) minced beef samples with normal (*A*) and high (*B*) contamination as in function of the storage time at 4 °C.

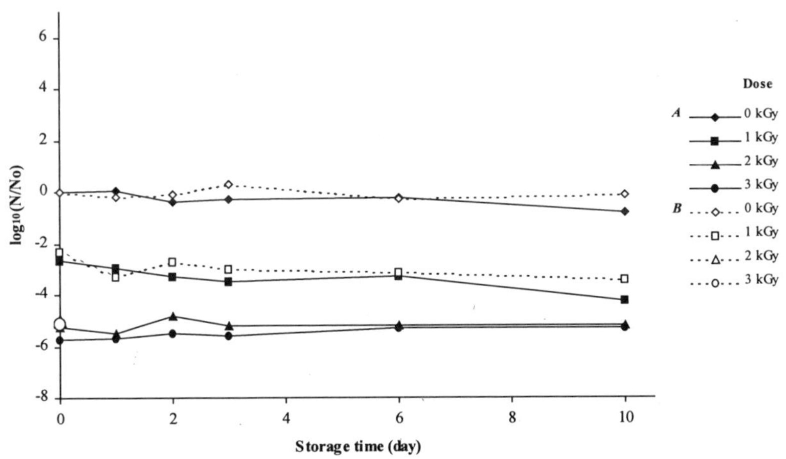

Fig. 7. Survival of *S. aureus* in aerobically packed unirradiated and irradiated (1, 2 and 3 kGy) minced beef samples with normal (*A*) and high (*B*) contamination as in function of the storage time at 4 °C.

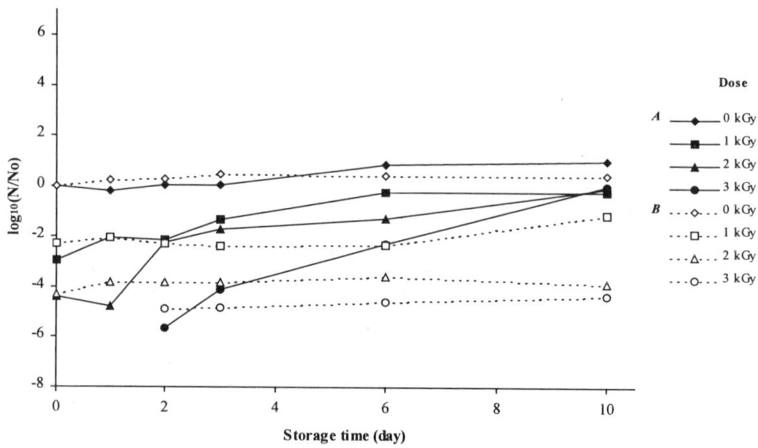

Fig. 8. Survival of *L. monocytogenes* in aerobically packed unirradiated and irradiated (1, 2 and 3 kGy) minced beef samples with normal (*A*) and high (*B*) contamination as in function of the storage time at 4 °C.

Control of Post-Harvest Loss of Grain, Fruits and Vegetables by Radiation Processing

P. THOMAS
Consultant, Food & Environmental Protection Section
International Atomic Energy Agency
Vienna, Austria

ABSTRACT

The feasibility for commercial application of radiation processing for enhancing food security by reducing post-harvest losses in agricultural products are discussed in this paper. It is reckoned that a sizeable quantity of world's food production is lost due to damage caused by bacteria, mould, insects and other pests. The major cause of post-harvest loss encountered in stored grains and other dried food crops is feeding by larvae and adults of several stored products insects. Many of the post-harvest chemical fumigants currently in use for control of insect infestation, such as ethylene dibromide (EDB), methyl bromide (MB) and ethylene oxide (ETO) are either banned or to be phased out because of their adverse impact on human health and environment. There are also reports of development by major grain pests of higher levels of resistance to phosphine, the other major grain fumigant used worldwide. Therefore satisfactory long term storage of staple crops may be at risk as the traditional fumigants are phased out and/or due to development of increased insect resistance. Practical control of insect pests, regardless of species or stage of development is possible by low dose irradiation in the dose range of 0.2 to 0.5 kGy. The advantages include short treatment time, no undesirable chemical residues in the food, no resistance developed by the insects and no significant changes in the physicochemical and functional properties or the nutritive value of the product.

Delaying the processes of ripening and over ripening in may fruits have been observed following treatment of the fruits in the physiologically mature but unripe state to low doses of ionizing radiation in the range of 0.25 to 1.0 kGy. More often, it is advantageous to combine irradiation with other post harvest procedures such as refrigeration, modified atmosphere and packaging to derive maximum benefits in terms of increased shelf-life. Combined treatments of radiation and hot water have been shown to provide good control of post harvest fungal rots and can be an effective alternative to chemical fungicides, some of which have been withdrawn in the recent times in view of their harmful effects on human health. In berry fruits such as strawberry, the mould free refrigerated shelf-life could be increased by irradiation at 2 kGy or by a combination of CO_2 and irradiation below 1 kGy.

Sprouting in tuber, bulb and root crops during storage is inhibited by irradiation at doses in the range of 0.02 to 0.15 kGy. The inhibition of sprouting is more effective when these crops are irradiated soon after harvest and curing when the commodities are in the dormancy period. However, it is important to control losses on account of microbiological spoilage by appropriate storage management such as aeration, temperature and humidity control of the storage environment. Large commercial scale irradiation of potatoes has been carried out in Japan since

1974. Since mid 1990's China has implemented large scale irradiation of garlic, up to 70,000 tons per annum in recent years, for sprout control. Pilot-scale studies in many countries have demonstrated the commercial feasibility of irradiation for sprout control on onions and yams under tropical storage conditions.

Radiation processing can thus provide an effective and safe alternative treatment for improving food security in many of the developing and less developed countries by reducing post-harvest losses of staple and non-staple crops when coupled with good handling, storage and transport practices.

INTRODUCTION

Reducing the wastage of crops after they are harvested is still a low priority for many countries in their efforts to bridge the widening gap between food production and food requirement. In most countries, especially the developing and less developed countries, the governmental as well as non-governmental efforts to meet the food shortage are essentially focussed on increasing agricultural productivity by improved farming practices and use of modern technological inputs. However, far less attention has been paid to reduce the post-harvest losses from the farm gate until it reaches the consumers house hold. Although as early as in 1974 the World Food Conference made a brief mention on the need for countries to endeavour to prevent wastage of food in all its forms [1], only in the very recent times the importance of post-production management and distribution systems has been recognised as a critical component in the global strategy for achieving food security and availability [2].

It is reckoned that a sizeable quantity of worlds food production is lost due to damage caused by bacteria, mould, insects and other pests. Conservative estimates of post-harvest losses indicate that average loss of cereals and legumes (pulses) is above 10%. For more perishable produce the losses exceed 20% of roots and tubers and 30% of fruits and vegetables [3]. The continued dependence on the use of chemicals in the post-harvest food loss prevention is increasingly coming under scrutiny in view of the harmful effects on human health and the environment. Many of the post-harvest fumigants currently in use such as ethylene dibromide (EDB), methyl bromide (MB) and ethylene oxide (ETO) are either banned or to be phased out because of their adverse impact on human health and environment. Therefore, satisfactory long term storage of staple crops may be at risk as these traditional fumigants are phased out. Research carried out world wide in the past four decades have shown that radiation processing is an effective and safe method for post-harvest loss reduction in agricultural products by controlling insect infestation in grains and other stored products, delaying the ripening and over ripening of fruits, and, preventing sprouting in bulb and tuber crops. The advantages of radiation processing and its commercial potential are discussed in more detail in the ensuing sections.

CONTROL OF INSECT INFESTATION IN GRAINS AND OTHER STORED FOOD PRODUCTS

The major cause of post-harvest loss encountered in stored grains and other food products such as cereals, legumes, and their products (flour, semolina), oil seeds, coffee beans, cocoa beans, dry fruits and nuts is feeding by larvae and adults of several stored product insects such as weevils, beetles and moths. The losses could reach enormous proportions under the warm ambient conditions of tropics and sub-tropics as these pests multiply very rapidly and can cause extensive product damage. It has been said that flour beetles are the "number one" pests to the food industry. The key basis for this belief is that population explosions are possible within a short period of time. If one assumes that each female can lay 400 eggs and that 60% of these eggs live to maturity, then in a four-month period under favourable environmental conditions, a population of almost 2 million can result [4]. Some of these insect pests feed inside the grain and whose immature stages develop there, and some feed externally.

Current methods of insect control in stored products

The various post-harvest procedures for the control of insects and mites in stored products are chemical control (fumigants, protectants, repellents), biological control (parasites and predators, insect pathogens, pheromones), low temperature storage, controlled or modified atmospheres (N_2 . CO_2), physical barriers and integrated pest management (IPM) using combination of these techniques [5]. The success and extent of using any one of these procedures would depend upon the nature of the product, storage infrastructure (silo, godown, warehouse, open storage under plastic covers), volume of product, and storage practice (as bulk grains or in packages such as sacks). Up to the 1990s many countries relied mainly on residual insecticides to control stored grain insects, but due to the increasing concern and objections to chemical residues in grain resulted in the use of fumigants , principally methyl bromide and phosphine. Presently, fumigation is the most widely used procedure that has served an important role in post-harvest pest control programs for many years. Fumigants are particularly useful for the control of stored product insects because they diffuse and penetrate relatively quickly into commodities. Methyl bromide is very effective against all stages of stored product pests, although the pupal stages of insects and the egg and hypopal stages of mites are usually hardest to kill. To control mites a dose 50% higher than that used for insects is required [5]. All of the fumigants used to control pest organisms are highly toxic to human beings, and they also may have adverse properties such as highly inflammable or corrosive nature and offensive odours, apart from their harmful residues in food materials.

Methyl bromide (MB), the most widely used post harvest fumigant for stored product insect control is to be phased out globally under the Montreal Protocol (an international treaty for the regulation of ozone depleting substances world wide under the auspices of the United nations Environmental Programme) in view of its strong potential as a chemical destroyer of the earth's protective ozone layer [6]. A phase-out by 2005 in Advanced Countries and by 2015 in Developing Countries has been agreed upon under this International Treaty in 1997. As the USA is the major producer of this chemical, it remains to be seen how other countries could obtain methyl bromide when there is no production in the USA after 2005 [7].

The other major post harvest fumigant used worldwide for the control of grain pests is phosphine in the form of aluminium phosphide tablets that release phosphine gas when exposed to moisture in the air or grain. Unlike MB, fumigation with phosphine require relatively long exposure periods (2 to 4 weeks) for achieving satisfactory insect kills. The development by major grain pests, such as the lesser grain borer (*Rhyzopertha dominica*), the rice weevil (*Sitophilus oryzae*), the rust-red flour beetle (*Tribolium castaneum*) and the flat grain beetle (*Cryptolestes ferrugineus*), of higher levels of resistance to the fumigant phosphine have been reported in China and India, and moderate resistance has been reported in Australia where it has been in use as the major grain fumigant This has necessitated in the application of increased dosages for obtaining good kill of resistant species [8]. As alternative protection and disinfestation methods including techniques involving carbon dioxide, nitrogen and refrigeration are generally much more expensive, many countries still rely on fumigation with phosphine.

Efficacy of radiation against insects

Research conducted world wide in the past four decades have shown that radiation processing of grains, grain products and other stored food products could be an effective alternative to the chemical treatment of foods. The dose required for insect control is fairly low, of the order of 1 kGy or less. The major difference between the action of fumigants such as methyl bromide and radiation process is that irradiated insects takes several days or weeks to die after exposure to the recommended doses for insect disinfestation. Delayed mortality of an insect is a function of irradiation dose and the susceptibility of the species and its developmental stages. The resistance of insect stages to radiation varies as development progresses from egg to adult [9]. Studies on the susceptibility to radiation of various species and stages of development show that practical control of insects, regardless of differences in species or stage of development, is possible in the dose range of 0.2 to 0.5 kGy. [10].Though lower doses do not cause immediate death of adults, low doses can prevent an increase in the population by lethal effects on immature stages, sterilization of adults and a reduction in adult longevity. Moreover these dosages can inflict damage to the mid gut of the insects thereby affecting their feeding ability, eventual starvation and death.[9, 11]. As a consequence very little product damage occur even if larvae and adult stages of the insect continue to survive for some period following irradiation. Studies have shown that the rate of food consumption of live adults of granary weevil (*Sitophilus granarius*) irradiated at a dose of 0.25 kGy decreased considerably [12] and that no grain damage was observed when wheat grains containing egg, larvae or pupae were irradiated to doses of 0.3 to 0.5 kGy [13]. To be economically feasible, the irradiation dose should be as low as possible and therefore the minimum dose to produce sterility or mortality of the insect should be used to achieve disinfestation [14]. It is pertinent to point out that phosphine, the other major fumigant currently in use for insect control also require long exposure times ranging from several days to weeks to kill the insects.

The advantages of radiation processing include short treatment time, no undesirable chemical residues in the food, no resistance developed by the insects and no significant changes in the physicochemical and functional properties or the nutritive value of the products. It has been reported that irradiation at insect disinfestation doses can in fact improve the nutritional quality of certain legumes by enhancing the conversion of flatulence causing oligosaccharides into the easily metabolizable simple sugars during germination [15]. In some developing countries marketing of grains and pulses and products like flour in consumer packs has shown a phenomenal growth in recent years. Insect infestation is often a serious problem during marketing of these products as insect eggs which are not visible to the human eye and present in the packaged product hatch and multiply rapidly under the warm ambient conditions leading to product damage and deterioration. Irradiation can effectively eliminate insect infestation in prepackaged products which is not feasible with fumigants.

Storage practices may vary from country to country. In many developing countries sizeable quantities of grains are stored as buffer stock or for public distribution either in jute bags or woven plastic bags. Since irradiation has no residual effect, it is necessary to prevent reinfestation by proper storage management and the use of insect resistant packaging materials for packaged products. In certain circumstances (e.g. long term storage, exposure to boring insects), it may be necessary to use insecticide, such as pyrethrin, impregnated packages [16].. Treatment with accelerated electrons is the cheapest for high volume operations [14] and may be the preferred method when large volumes are to be treated for storage in insect proof silos. Commercial irradiation of grain with electron accelerators was used in the Ukraine during 1980s [17]. In dried fruits and nuts postharvest insect infestation cause losses through direct damage and product contamination and by creating favourable conditions for mold growth and product degradation. The prospects for irradiation control of insect pests of dried fruits and walnuts has been addressed recently [18].

POST-HARVEST LOSS REDUCTION IN FRUITS AND VEGETABLES

Control of fruit ripening

Biochemical and physiological changes leading to ripening in fruits make them more susceptible to mechanical damage while handling and transporting as well as to the attack of rot producing fungal pathogens during storage and marketing. Storage temperature plays a very critical role in the rate of metabolism in fresh fruits and vegetables. In these commodities extension of the short shelf-life is highly desirable, and in some cases, critical for their transport and marketing. Refrigeration, controlled or modified atmosphere storage and removal of ethylene from the atmosphere surrounding the produce can delay the ripening process, but has limitations with many of the tropical fruits due to their susceptibiltiy to chilling injury or carbon dioxide injury when exposed for long duration. Delaying the processes of ripening and over ripening in fruits such as bananas, mangoes, papayas, and other tropical, subtropical and temperate fruits have been observed following exposure of fruits in the physiologically mature but unripe state to low doses of ionizing radiation in the range of 0.25 to 1.0 kGy [19, 20-22]. In many of the tropical fruits, the optimal dose for delaying the ripening process and the maximum dose that can be tolerated with out causing phytotoxicity may vary between fruits and among varieties of the same fruit.

Combination processes involving irradiation and other physical treatments

It is often advantageous to combine irradiation with other post harvest procedures such as refrigeration, modified atmosphere and packaging to derive the maximum benefit in terms of increased shelf-life. Thus, irradiation has been shown to increase the shelf-life of some tropical fruits such as bananas and mangoes at non chilling temperatures and their transport over long distances [20]. Presently trade in tropical and subtropical fruits from developing countries to markets in industrially advanced countries is mostly by the costly air transport on account of their perishable nature and short shelf-life. Irradiation processing may make it possible to transport such produce by sea freight in refrigerated cargoes. Also, in combination with other physical treatments such as hot water dip at temperatures of 50 - 55°C for 5 - 10 min, low dose irradiation can effectively control post harvest fungal rots. Such combination treatments have been shown to provide good control of post-harvest fungal diseases as well as shelf-life enhancement in many fruits including mangoes and papayas.[20, 23]. Some of the fungicides that have been in use for applications related to post-harvest control of fruit rots have been withdrawn in the recent times in view of their harmful effects on human health. Combination treatments involving irradiation could thus offer an effective alternative to chemical fungicides in the control of post-harvest diseases during storage, transport and marketing of fruits.

In berry fruits such as strawberries, the mould free refrigerated shelf-life was increased by irradiation at 2 kGy [22] or by a combination of CO_2 and irradiation below 1 kGy [14]. Small scale commercial application of irradiation to control spoilage losses of strawberries has been carried out increasingly in the USA since 1992. In this context, it is worth mentioning that irradiation at the dose levels used for delay of ripening could also provide quarantine security against fruit flies and other pests of fruit crops entering international trade (in this Conference).

In general, irradiation was found to provide no special advantage in the preservation of most fresh vegetables. However, recent research have shown that both shelf-life and microbial safety of fresh cut or minimally processed refrigerated prepackaged vegetables and salad vegetables such as shredded carrots, carrot cubes and bell peppers can be improved by irradiation in the dose range of 1-2 kGy [24-26,]. Combination of chlorine and irradiation at 0.5 kGy significantly reduced microbial population and shelf-life of fresh cut chopped lettuce [27] and shredded carrots in modified atmosphere packaging [28].

INHIBITION OF SPROUTING IN STORED BULB, TUBER AND ROOT CROPS

Sprouting of bulb, tuber and root crops during storage can lead to both qualitative and quantitative deterioration due to excessive shrivelling, loss of nutritive value and marketability. Storage at low temperatures by circulating the ambient cold air or by mechanical refrigeration with or without the use of chemical sprout inhibitors is practised in some countries to control sprouting in storage. The storage temperature may vary from 2-4 °C to 7-10 °C depending on the end use. Potatoes meant for industrial processing into chips and french fries are usually stored at the higher temperature coupled with chemical sprout suppressants to prevent sprouting occurring at such storage temperatures. Potatoes meant for table purposes are stored at temperatures of 2-4 °C for periods up to 6 months using mechanical refrigeration, but removal to ambient conditions for marketing purposes usually results in profuse sprouting in such tubers. Chemical sprout inhibitors such as maleic hydrazide (applied as pre-harvest spray in onions and potatoes), propham and chlorpropham (post-harvest application for potatoes) used under cool temperate climates for the control of sprouting in onions and potatoes are not very effective under the warm tropical and subtropical ambient conditions. Moreover, these chemicals leave residues considered by some to be harmful, and many countries have prohibited their use [17]. Yam, a staple carbohydrate food crop in the yam zone of West Africa and also of importance in Southeast Asia, regions of the Pacific, and the Caribbean are sensitive to low-temperature injury and therefore not feasible to be stored for long durations under refrigeration. Similarly, chemical sprout inhibitors were also reported to be generally not very effective in inhibiting sprouting during normal ambient storage [29].

Efficacy of radiation for sprout inhibition

Sprouting of crops such as onions, shallots, garlic, potato and yam during storage is inhibited by irradiation at doses in the range of 0.02 - 0.15 kGy. The lowest effective dose varies among the different crops and also between cultivars of the same crop. Inhibition of sprouting is more effective when these crops are irradiated soon after harvest and curing when the commodities are still in the dormant state [30-32].

It is to be emphasized that the objective of treating harvested bulbs, tubers and root crops with low dose ionizing radiation is to prevent the physiological processes leading to sprouting and consequent product deterioration during extended storage. Losses on account of other causes such as microbiological spoilage can continue to occur in storage and needs to be controlled by appropriate storage management procedures such as proper aeration, temperature and humidity control of the storage environment. For example, extended storage of potatoes may not be feasible under warm tropical ambient climates, as bacterial and fungal rots may take a heavy toll of the stored produce. Therefore, storage at cooler temperatures of 10 - 20 °C is recommended depending on the intended storage duration The feasibilty of irradiation for reduction of sprouting losses in potatoes and onions under cooler or well aerated ambient conditions has been demonstrated in pilot scale storage trials [33-35]. Large commercial scale irradiation of potatoes has been carried out in Japan since 1974. Since mid 1990s China has implemented large scale irradiation of garlic for sprout control. It is reported that during 1999 for the period up to september over 70,000 tons of garlic has been irradiated for sprout control [Zhu Jiang, personal communication]. Recent studies in Cote D'Ivoire and Ghana as part of an IAEA Co-ordinated Research programme have demonstrated the technological feasibility and commercial potential of irradiation for control of sprouting in yams under traditional storage conditions [36].

Irradiated tuber and bulb crops have been shown to be suitable for further industrial processing e.g. french fries from potatoes and dehydrated flakes from onions and garlic [31,32, 37]. It has been reported that a large percentage of the 15-22,000 tons of potatoes irradiated annually since 1974 by the Shihoro Agricultural Co-operative Association at Shihoro in Hokkaido, Japan is used for processing into chips, shoe strings, frozen french fries and other products towards the end of the storage season when potatoes harvested from other growing areas are not suitable for processing purposes [37].

OUTLOOK FOR RADIATION PROCESSING

Irradiation offers a sustainable, less energy intensive and environment friendly broad spectrum treatment for the control of a wide variety of insect pests, spoilage causing organisms and physiological processes in many durable and perishable crops. Pilot-scale storage studies in many countries have pointed to the commercial potential and economic viabilty of irradiation as a post-harvest process for specific food items. Based on these pilot studies, some countries are in the process of establishing commercial facilities for the treatment of bulb and tuber crops and grains. The acceptance of irradiation as a quarantine treatment for horticultural products entering trade may also indirectly contribute to reduction in post harvest losses by improving the market life of these products. National regulations in many countries, and international (Codex) standards are in place and public and food industry perceptions on the safety and benefits of the radiation process are more positive. In view of the phasing out of methyl bromide, the widely used post harvest fumigant and development of increased resistance by insect pests of stored grains to phosphine, irradiation can provide an effective and safe alternative treatment for improving food security in many of the developing and less developed countries by reducing post-harvest losses of staple and non-staple crops when coupled with good handling, storage and transport practices.

REFERENCES

[1] UN., Universal declaration on the evaluation of hunger and malnutrition. Report of the World Food Conference, United Nations, New York (1975).

[2] FAO., Food for consumers: Marketing, processing and distribution. World Food Summit technical background document. Food and Agriculture Organization., Rome, Italy (1996).

[3] Anon. 1998. Safeguarding Our Harvests, International Consultative Group on Food Irradiation, Vienna, Austria.

[4] Baur, F.J., 1992. *Insect Management for Food Storage and Processing*, American Association of Cereal Chemists, Inc., St.Paul, Minnesota, USA.

[5] Bond, E.J., 1992. Fumigation of raw and processed commodities, in *Insect Management for Storage and Processing*, American Association of Cereal Chemists, Inc., St. Paul, Minnesota, USA, pp 143-147.

[6] UNEP, 1992. Fourth meeting of the parties to the Montreal Protocol on substances that deplete the ozone layer, United Nations Environment Programme, Copenhagen.

[7] Loaharanu, P., 1998. International developments of food irradiation, Food Irradiation 99- The Solution to the Food Safety Problem, May 12-14, 1999, Washington, D.C., USA.

[8] Lehane, R., 1999. Joining a new battle with grain pests, Partners in Research for Development, No.12: 26-30.

[9] Tilton, E.W. and Brower, J.H., 1983. Radiation effects on arthropods, in *Preservation of Food by Ionizing Radiation* Vol.II (eds. E.S. Josephson and M.S. Peterson), CRC Press, Boca Raton, Florida, USA., pp 269-316.

[10] Watters, F.L., 1992. Potential of ionizing radiation for insect control in the cereal food industry, in *Insect Management for Food Storage and Processing,* American Association of Cereal Chemists, Inc., St.Paul, Minnesota, USA, pp 265-274.

[11] Tilton, E.W. and Burditt, A.K, Jr., 1983. Insect disinfestation of grain and fruit, in *Preservation of Food by Ionizing Radiation* Vol.III (eds. E.S.Josephson and M.S.Peterson), CRC Press, Boca Raton, Florida, USA.

[12] Zewar, M.M. and Abdel-Salam, C.A., 1988. Food consumption by irradiated *Sitophilus granarius* (L.) and *Rhyzopertha dominica* F., Agricultural Research Review 66: 55-61.

[13] Aldryhim, Y.N. and Adam, E.E., 1999. Efficacy of gamma irradiation against *Sitophilus granarius*, J. Stored Products Res., 35: 225-232.

[14] WHO., 1994. *Safety and Nutritional Adequacy of Irradiated Food,* World Health Organization, Geneva.
[15] Machaiah, J.P., Pednekar, M.D. and Thomas, P., 1999. Reduction in flatulence factors in mung beans (*Vigna radiata*) using low-dose gamma irradiation, J. Sci. Food Agric.,79: 648 -652.
[16] Anon., 1991. *Insect Disinfestation of Food and Agricultural Products by Irradiation,* International Atomic energy Agency, Vienna, Austria.
[17] Diehl, J.F., 1995. *Safety of Irradiated Food,* Marcel Dekker, New York.
[18] Johnson, J. and Marcotte, M., 1999. Irradiation control of insect pests of dried fruits and walnuts, Food Technol., 53 (6): 46-51.
[19] Moy, J.H., 1983. Radurization and radicidation: fruits and vegetables, in *Preservation of Food by Ionizing Radiation* Vol.III, CRC Press, Boca Raton, Florida, USA, pp 83-108.
[20] Thomas, P., 1986. Radiation preservation of foods of plant origin. 3. Tropical fruits: Bananas, mangoes and papayas, CRC Critical Reviews in Food Sci.Nutr., 23: 147-205.
[21] Thomas, P., 1986. Radiation preservation of foods of plant origin. 4. Sub-tropical fruits: Avocados, citrus and grapes, CRC Critical Reviews in Food Sci.Nutr., 24: 53-89.
[22] Thomas, P., 1986. Radiation preservation of foods of plant origin. 5. Temperate fruits: Pome fruits, stone fruits and berries, CRC Critical Reviews in Food Sci. Nutr.24: 357400.
[23] Brodrick, H.T. and van der Linde, H.J., 1981. Technological feasibility studies on combination treatments for subtropical fruits, in *Combination Processes in Food Irradiation,* International Atomic Energy Agency, Vienna, Austria, pp 141-156.
[24] Farkas, J., Saray, T., Mohacsi-Farkas,C., Horti, K. and Andrassy, E., 1997. Effects of low dose gamma irradiation on shelf-life and microbiological safety of pre-cut/prepared vegetables, Adv. Food Sci. (CMTL), 19: 111-119.
[25] Ramamurthy, M.S., Deshpande, A., Alur, M.D. and Thomas, P., 1998. Improvement of shelf-life and microbilogical quality of minimally processed refrigerated capsicum by gamma irradiation, 4th International Convention of Food Scientists and Technologists (IFCON-98), Mysore, India.
[26] Chervin, C. and Boisseau, P., 1994. Quality maintenance of "ready to eat" shredded carrots by gamma irradiation, J.Food Sci. 59: 359-361, 365.
[27] Hagenmaier, R.D. and Baker, R.A., 1997. Low-dose irradiation of cut iceberg lettuce in modified atmosphere packaging, J.Agric. Food Chem. 45: 2864-2868.
[28] Hagenmaier, R.D. and Baker, R.A., 1998. Microbial population of shredded carrot in modified atmosphere packaging as related to irradiation treatment, J.Food Sci. 63:162-164.
[29] Adesuyi, S.A. and Mackenzie, J.A., 1973. The inhibition of sprouting in stored yams, *Dioscorea rotundata* Poir. By gamma irradiation and chemicals, in *Radiation Preservation of Food,* International Atomic Energy Agency, Vienna, Austria, pp 127-136.
[30] Matsuyama, A. and Umeda, K., 1983. Sprout inhibition in tubers and bulbs, in *Preservation of Food by Ionizing Radiation,* (eds.) E.S.Josephson and M.S.Peterson, CRC Press, Boca Raton, Florida, USA, pp 159-214.
[31] Thomas, P., 1984. Radiation preservation of foods of plant origin.1.Potatoes and other tuber crops, CRC Critical Reviews in Food Sci. Nutr., 19: 327-379.
[32] Thomas, P., 1984. Radiation preservation of foods of plant origin. 2. Onions and other bulb crops, CRC Critical Reviews in Food Sci.Nutr., 21: 95-136.

[33] Matin, M.A., Bhuiya, A.D., Amin, M.R., Malek, M.A., Nahar, G., Saha,A.K., Rahaman, S., Khatoon, J.,Shajahan, R. Md., and Siddiqui, A.K., 1996. Irradiation of onions, pulses and dried fish: Process control, storage, Testmarketing and economic analysis of the process, in *Food irradiation with emphasis on process control and acceptance in Asia*, IAEA-TEC-DOC-871, International Atomic Energy Agency, Vienna, Austria, pp 19-50.

[34] Sattar, A., 1996. Irradiation preservation of onions, potatoes, spices and poultry meat, in *ibid*, pp 51-68.

[35] Thomas, P., Padwal-Desai, S.R., Srirangarajan, A.N., Joshi, M.R., Janave, M.T., Bhonde, S.R. and Qadri, S.M.H., 1986. Pilot scale storage tests on the efficacy of gamma radiation for sprout inhibition of onions under commercial conditions, J.Food Sci. Technol. 23(2): 79-84.

[36] Report of the Final FAO/IAEA Research Co-ordination Meeting (RCM) on "Study of the Impact of Irradiation to Prevent Food Losses in Africa", September 1999, International Atomic Energy Agency, Vienna, Austria.

[37] Anon., 1997. *Irradiation of bulbs and tuber crops- A compilation of technical data for its authorization and control*, International Atomic Energy Agency, Vienna, Austria.

Modern Trends in Post-Production Technology

M. SATIN
Chief, Agro-Industries and Post-Harvest Management Service
Food and Agriculture Organization of the United Nations

ABSTRACT

Global consumer demand for fresh tasting and nutritious, high-quality foods has created considerable interest in the development of new or improved postharvest storage and food-processing techniques. The competitive struggle for markets, which has resulted from more liberalised trade regimes, requires a much greater emphasis on highly efficient and effective post-harvest technologies including handling, processing and distribution in order to access markets further and further afield. While there is no doubt that traditional methods have contributed to improved food availability and safety in the past, conventional heating and cooling reduce many quality attributes of foods.

Consumers are demanding access to more and more fresh products, no matter where in the world they are. This demand, together with the banning of ozone-depleting fumigants has created the incentive to ensure a much wider use of improved post-harvest storage methods as well as new and emerging non-thermal technologies.

Several factors lead to the cumulative causes of post-harvest food losses in developing countries. These include inefficient harvesting and handling methods, poor processing techniques, inadequate methods of storage and distribution and even poor preparation of foods in the home. The wide range of current technologies for post-harvest anti-microbial or disinfestation treatment are described including hot water application, treatment with synthetic as well as natural compounds and physical treatments. The most promising of the new and emerging post-harvest technologies are also described, including food irradiation, high-pressure processing, pulsed-electric fields and pulsed light applications. Despite the demonstrated safety and value of the new technologies, concern still remains regarding the misperceptions related to their application.

INTRODUCTION

The global consumer demand for high-quality foods that are both fresh tasting and nutritious has created considerable interest and investment in the development of new or improved post-harvest storage and food-processing techniques. The competitive struggle for markets, which has resulted from more liberalized trade regimes, has required a much greater emphasis on efficient and effective post-harvest handling, processing and distribution to access markets further and further afield. Traditional post-harvest food-processing technologies such as freezing and canning are no longer at the forefront of consumer demand. While there is no doubt that these methods have contributed to improved food availability and safety in the past, conventional heating and cooling reduce many quality attributes of foods. Consumers are demanding access to more and more fresh products, no matter where in the world they happen to be located. This demand, together with the banning of ozone-depleting fumigants, has generated the incentive to ensure a much wider use of improved post-harvest storage methods as well as new and emerging non-thermal technologies.

Several factors lead to the cumulative causes of post-harvest food losses in developing countries. These include inefficient harvesting and handling methods, poor processing techniques, inadequate methods of storage and distribution and even poor preparation of foods in the home. Losses are, of course, not simply calculated as absolute volume losses, but quality losses as well.

Traditional marketing systems can contribute to reduced returns to farmers by involving several changes of hands before the produce finally reaches the consumer. Inadequate storage and poor road and transport systems make distribution to key points extremely difficult, particularly with perishable horticultural crops during critical seasons.

The actual causes of volume and quality losses can be grouped into two main categories, the primary and secondary causes of post-harvest losses.

Primary Causes

The primary causes of loss are those that directly affect the food. They may be classified into the following sub-groups:

(a) **Biological:** The direct consumption of food by insects, rodents or birds; aside from material losses, the level of contamination by excreta, hair, webbing and odours can be so high that the food will be condemned for human consumption.

(b) **Microbiological:** The damage of stored food by fungi and bacteria; micro-organisms can damage goods to the point where they become unacceptable because of rotting or other defects, such as aflatoxin, produced by moulds or unsightly bacterial rotting.

(c) **Chemicals:** Many of the natural chemical constituents present in foods can react causing loss of color, flavor, texture and nutritional value; examples include the reactions that cause browning in dried fruits.

(d) **Biochemical reactions:** A large number of natural enzyme-activated reactions can occur in foods during storage, giving rise to off-flavors, discoloration and softening.

(g) **Physiological:** Natural respiratory losses which occur in all organisms account for a significant level of weight loss and can also generate damaging heat; changes during maturation, such as ripening, wilting and sprouting, also increase the susceptibility of the product to mechanical damage or infection by pathogens; a reduction in nutrient level, such as vitamin C, and consumer acceptability can also accompany these changes.

Of the above primary causes, the biological, microbiological, mechanical and physiological factors are responsible for the majority of the losses in perishable crops.

Secondary Causes

While primary causes do the actual damage, the secondary causes are those that encourage the conditions which result in the primary cause of loss. They are usually the result of inadequate training, inadequate or non-existent storage structures, unsuitable technologies and ineffective quality control.

Magnitude of Losses

Post-harvest losses are difficult to determine accurately. Depending upon the year and the country, it is possible to find individual cases with losses ranging anywhere from near zero up to one hundred per cent. This high variability is dependent upon a number of conditions. Staple foods such as cereal grains can be stored in good condition for several years, whereas perishable foods such as fruits and vegetables spoil quickly, unless given particular treatment such as controlled-atmosphere storage, canning, freezing or irradiation. Generally speaking, horticultural crops suffer from higher losses than cereals.

The pattern of losses varies widely from country to country. There is a marked contrast between the site of major losses in developed countries and those of developing countries. In a typical developed country, losses can be fairly high during harvesting because the agricultural machinery that is used to harvest the crops leaves a good deal of the production in the field and mechanically damages some of it. Considerable quantities of foods may be discarded at the point of harvest because they are the wrong size, shape or colour. But these are planned losses. In developing countries, harvesting losses are usually lower because most of the crop is handpicked. The amount of material rejected in the developing countries is less because food is such a precious commodity and the expectations of quality and uniformity are generally lower than those in developed countries.

Modern Post-Harvest Technologies

Chemical and physical post-harvest technologies are chiefly used to reduce decay and physiological irregularities and to delay normal senescence metabolism.

Physico-Chemical Post-Harvest Treatments to Control Spoilage

Conventional post-harvest treatments by dipping, drenching or spraying are effective, but can be significantly influenced by the commodity and variety of spoilage organisms such as moulds. Naturally, commodities differ in their resistance to moulds as measured by the effectiveness of various anti-fungal compounds, as well as the effects of their specific morphologies (calcium content, waxy layers of the skin, etc.).

In addition to conventional chemicals such as benzimidazoles, thiazoles, phenols and sulfur, natural plant extracts are gaining increased attention and application. Products such as thymol, cinnamaldehyde, salicylaldehyde and chitosan or a mixture of extracts are currently being evaluated and others are already being marketed.

For some commodities, hot water treatment at 55°C for 2-7 minutes, or 42°C for 30-60 minutes, reduces disease appearance. Temperature-time regimes are commodity-specific and operate within narrow margins. The combination of irradiation and heat treatment is also being currently investigated.

Post-Harvest Cooling

The competitive need to market products of higher quality and to extend shelf life to allow trade in more distant markets requires rapid cooling after harvest, especially for berries and small fruits. All delays in the application of cooling will shorten shelf life. In addition to conventional cooling rooms, pre-cooling using vacuum and high velocity forced-air systems is also available. In developing countries, storage techniques are generally limited to mechanical cooling, forced air cooling or storage in naturally cool caves. These technologies can be complemented or optimized by proper ventilation or simple modified atmosphere systems. To preserve fruit and vegetables in developing countries, research on solar drying, lactic acid fermentation, fresh packing of fruit with added tropical berry juice containing high ascorbic acid levels and low pH, or even natural anti-browning agents, such as grapefruit or avocado seeds, is being encouraged.

In developed countries, fairly sophisticated storage techniques exist. Because of the rapidly increasing markets for horticultural products, requirements for compressor capacity have substantially increased. New, non-ozone depleting refrigerants are increasingly employed and a far greater use of modified or controlled atmospheres, ultra-low oxygen environments and ethylene rooms is now evident.

Fresh pack and minimally processed products are developing very quickly. Because of the nutritional and quality changes that result from traditional processing, a significant amount of research has been carried out to develop new methods of treating foods. Some of them are already commercial, while others still require fine-tuning to make them competitive with technologies currently being used. Although the destruction or inhibition of unwanted spoilage and disease organisms is the common goal of these technologies, the resultant end-products can be divided into two types. The first of these products is heat-processed and does not resemble natural, raw products in any way. The major advance is in the way the thermal treatment is carried out. The second type of process seeks to destroy or eliminate unwanted micro-organisms without significantly changing the basic character of the food, so that the consumer can utilise it as a fresh product but without the fear of food-borne disease.

New and Emerging Food Technologies

The most promising among these emerging technologies are food irradiation and high-pressure processing, followed by pulsed-electric fields and pulsed light applications. Although it is beyond the scope of this paper to cover these technologies in detail, a brief review of their applications in food processing is in order.

Food Irradiation

Food irradiation is a century-old process where foods are exposed to a highly penetrating form of energy - gamma rays or high-energy electrons. Because these forms of energy can fairly evenly penetrate solid foods, gamma rays and high-energy electrons can uniformly inactivate the DNA of unwanted micro-organisms without changing the basic nature of the food itself. Fresh irradiated foods with exceptional hygienic and eating qualities are very similar to fresh, untreated foods, except for the label and, in some cases, an improved appearance.

Food irradiation can be used on most fruits and vegetables, as well as meat, poultry, fish, shellfish and seafood, spices, potatoes, grains and a host of other commodities. Depending upon the food product and treatment in question, the cost of food irradiation varies between ½ -5 cents per pound. Considering the benefits it provides to the practical and hygienic qualities of foods, this is a very reasonable price to pay for the benefits it brings to consumers. The use of irradiation as a quarantine treatment is well established and it is expected to get much more widespread commercial use in future. Of particular interest to this area is the recent work on low energy 'soft-electrons' currently being carried out at the National Food Research Institute of Japan.

A review of the most recent studies carried out by universities and national polling organizations has indicated that current consumer fears of food-borne disease far exceeds any fears of food irradiation. Consumers will willingly buy irradiated food, particularly when they are aware of the improved hygienic quality of the food. Since food irradiation is the only physical process that requires clear labeling, these products will be apparent on the marketplace. Access to irradiated foods is therefore not a problem of consumer resistance - it is really a problem of industry resistance. The biggest barrier is the supermarket industry, which feels that consumers will not accept irradiated foods; however, this statement is not supported by the published facts.

High Pressure Processing

Another method which is designed to maintain foods in their raw, natural state and only destroy the contaminating organisms is high-pressure processing. When organisms are placed under conditions of very high pressure, their proteins will be denatured, resulting in the loss of enzymatic and biological activity. The organisms lose their viability and perish. This is much more easily accomplished with vegetative bacteria than with spores, which require much higher pressures. This technique is not new and was first applied a full century ago. In 1899, pressures in the range of 5,000 to 7,000 kg/cm^2 were used to reduce microbial levels in milk and meats. Test results showed that a five- or six-log-cycle reduction in bacterial count was possible in milk when it was subjected to high pressures.

Other studies showed that high-pressure processing was beneficial in extending the shelf life of processed fruits. These early studies demonstrated that the application of high pressure had effects similar to the use of high temperature on proteins and microbial population in foods, but, unlike thermal treatments, was able to maintain the fresh character of the original products.

Modern high-pressure treatment involves subjecting food materials to pressures as high as 9,000 atmospheres applied uniformly throughout the food, independent of its mass. This method is currently being applied to fruit products such as jams and juices, particularly in Japan. Raw materials are placed in chambers under very high pressure for 10-60 minutes prior to opening the release valve. Both the equipment and the process are costly and, as a result, the major commercial limitation to this method is economic rather than technical. As a commercial process, high-pressure treatment can significantly increase the cost of foods. This has severely limited the wider adoption of this process and will probably continue to do so for some time to come. Examples are shown in Table 1.

Table 1. Some examples of products processed using high-pressure technology and changes in quality attributes other than microbial changes.

Product	**Process and Quality Attributes**
Avocado Puree	Prevents discoloration. Inhibition of undesirable browning reactions in presence of low pH.
Banana Puree	Prevents discoloration. Reduction in Polyphenoloxidase activity when combined with blanching.
Black Beans	Cooking. Higher water absorption and less cooking time.
Cheese	Precise control of rennet coagulation of milk.
Jam	Commercial Production (Meiji-ya, Japan). Improved retention of color and flavor of fresh fruit.
Meats	Thawing. Reduction in drip loss and less color change.
Meats, tenderised	Commercial production (Fuji Chiku and Mutterham, Japan). Improved retention of sensory characteristics.
Orange juice, fresh-squeezed	Preservation. Retention of colour and cloud stability during storage.
Pink grapefruit juice, fresh-squeezed	Preservation. Retention of color and cloud stability during storage.
Pork sausage	Manufacturing. Moister, denser and more tender sausages with more retention of color than if heat-treated.
Potato	Freezing. Reduction in freezing time in potato cylinders.
Rice paste with herbs (*Yomogimochi*)	Commercial Production (Japan). More desirable sensory properties than if heat-treated.
Soya proteins	Manufacturing. Less firm but more elastic and extensible gels. Improved preservation of color and initial aroma.
Surimi	Control of enzyme activity. Enhanced activity of transglutaminase in surimi with increased gel strength.
Surimi, Pacific Whiting	Gelation. Increased gel strength in surimi.
Tofu	Freezing. Production of small-size ice crystals.

Tomato juice	Juice production. Modification of physical and sensory characteristics deemed desirable.
Yoghurt	Storage. Reduced syneresis.

High Electric Field Pulse (HEFP) Treatment

This process seeks to destroy unwanted bacterial cells without the application of heat, in order to have a product remain in its raw natural state but without viable contaminating organisms. Similar to high-pressure processing, the beneficial effects of pulsed electric fields on inactivating microbial levels in foods have been known for several decades. Indeed, the pasteurizing effects of electric fields in foods were first observed as far back as the early 1900s. During the 1960s, it was realized that the process appeared to create pores in microbial cell membranes by subjecting cells to high voltage. If the applied electrical field is greater than a certain critical value, then the cell wall ruptures. Although the entire mechanism of cell membrane breakdown in electric fields is not fully understood, the process is used regularly in creating pores in a cell membrane. The technique is known as electroporation, and it has been used in the field of biotechnology to introduce foreign DNA into microbial cells.

Recently, an HEFP pilot unit integrated with an aseptic packaging machine for processing of fresh orange juice on a pilot scale has been described. The authors concluded that the treatment inactivated 99.9 per cent of microbial flora. Compared with heat pasteurization, the HEFP-treated orange juice retained more vitamin C and flavor.

Since the utilization of this technology for the treatment of food is still in the experimental stages, there are many unsolved problems. One problem is that the process is not particularly selective and intact cells of the food are also damaged. This may lead to changes in the texture of those foods, such as a loss of crispness in carrots, for instance. The fact that low levels of treatment may result in minor injury that is reversible begs the question as to whether microbial cells will be able to develop resistance to increasing levels of treatment before irreversible damage is achieved. Another concern is that most bacterial spores are very resistant to this treatment. Since spores are a particular problem in natural plant products, the applicability of this technology to fruits and vegetables will be severely limited if this problem cannot be overcome through the employment of other combination treatments. At the present time, this process is not in commercial use, but it may one day become a reality.

Table 2. Examples of foods processed using pulsed-electric fields.

Product	**Process and Quality Attributes**
Apple Juice, fresh and reconstituted	Pasteurization. No change in solids concentration, pH and vitamin C. Loss of calcium, magnesium, sodium and potassium. No sensory differences between processed and untreated juices.

Commercial cheese sauce, reformulated	Preservation. Better flavor and appearance than comparable products.
Green Pea Soup	Cooking. No sensory difference after 4 weeks storage at 4°C.
Liquid Whole Egg	Pasteurization. Prevention of coagulation, superior quality.
Orange juice	Preservation at pilot-scale. Less than 6% flavor loss, negligible vitamin C and color change.
Orange juice, fresh	Pasteurization. Minimal loss of flavor, color and vitamin C.
Salsa	Preservation. Better flavor and appearance.
Spaghetti Sauce	Aseptic processing. Acceptable after 2 years and 80°F storage.

Ohmic Heating

In order to improve the uniform transfer of heat to food products, attempts were made to carry out the thermal treatment in scraped-surface heat exchangers, followed by aseptic packaging in aluminum cans. This worked well for uniform fluid products such as smooth puddings, but failed when larger particulate matter was incorporated. This was simply because the particulate material heated up more slowly than the surrounding fluid. As a result, the entire process had to be slowed down to ensure that the particulates were properly treated. One solution is ohmic heating.

Ohmic heating itself is not a new concept. An electric current is passed through the food and the solid materials heat up as fast or even faster than the surrounding fluid. The electrical current heats up all the particulates quickly and efficiently so that the final temperatures required to kill or inhibit bacteria are achieved with far less total cooking. This results in products with improved quality. However, although the process is now commercial, the total volume of products treated in this manner is thought to be very small.

The two last methods I would like to mention are pulsed light treatment and oscillating magnetic fields. Neither method has found commercial application in post-harvest horticulture so far, but may in the future.

Pulsed-Light Treatment

Pulsed-light treatment involves the use of high-intensity light for the purpose of killing micro-organisms on the surface of food or packaging materials. This procedure, developed under the trade name PureBright, uses a light spectrum containing wavelengths all the way from ultraviolet to near-infrared. The light spectrum generated with this equipment is similar to that of the sunlight reaching the earth's surface but the intensity of PureBright is 20,000 times higher. The intense flashes of light produced by the PureBright system are used to destroy micro-organisms.

Results indicate that application of pulsed light can reduce up to 9 logs of vegetative micro-organisms and more than 7 logs of bacterial spores on smooth, nonporous surfaces, such as those of packaging materials. When the surfaces are more complex and porous, such as in the case of food materials, then the microbial reduction is only 2 to 3 log cycles. The reported costs of equipment amortization, lamp replacement, electricity and maintenance indicate expenditures of only a few tenths of a cent per square foot of treated area.

OSCILLATING MAGNETIC FIELDS

Published studies in the literature show mixed results in the inhibition of micro-organisms when placed in oscillating magnetic fields (OMF). Some studies indicate that magnetic fields have an inhibitory effect on the microbial population, while others show no effect or, in some cases, even a stimulating effect. Mechanisms describing these observations are under scientific inquiry. In one study, foods with high electrical resistivity were subjected to one or more pulses of oscillating magnetic fields. It was observed that a single pulse of magnetic field generally decreased the microbial population by at least two orders of magnitude. The process involves little thermal energy input, thus avoiding thermal denaturation of food components during treatment. However, the method is still experimental and more research is needed to understand the changes in microbial population when treated with OMF.

Minimally-Processed Foods

While the above methods were developed to kill the micro-organisms present in food, many other methods, some of them thousands of years old, are employed to simply control their excessive growth. Since the minimum infective dose of pathogens is fairly high for many food-borne diseases, these methods have played a very important role in maintaining food safety over the centuries. Common examples of these methods are acidification (as in pickled vegetables), fermentation, the use of high concentrations of salt or sugar, freezing or cooling and drying. These methods continue to be very popular in the manufacture of traditional food products. In fact, this approach has been broadened into a new class of foods, described as *minimally-processed foods*.

In order to increase the eating quality of contemporary ready-to-eat processed foods, a battery of gentle post-harvest techniques has been combined into fully integrated treatments. The philosophy behind this approach is based upon the notion that each technique alone might not destroy micro-organisms, but when employed together they are very effective. The concept of minimal processing theoretically involves the care of foods throughout the entire post-harvest system - from the farm-gate all the way to the consumer. This minimal processing approach is also called *hurdle technology*, simply because a series of hurdles is placed in the way of the micro-organisms' growth and survival; e.g. combinations of weak acid treatments with modified atmosphere packaging, or mild heating with reduced water activity, or alternative doses of mild heating and chilling. Indeed, the preparation and marketing of fresh-bagged salad in modified atmosphere packs has been a remarkable commercial success.

The marketing concept for minimally-processed foods is based upon the perceived consumer desire for more natural, less processed, high quality home-made-style food preparations. Chilled ready-to-eat or ready-to-heat foods are a very rapidly growing segment of the market. Some of the best examples of this are the so-called *sous-vide* or vacuum-packed products. These products are prepared under very rigorous hygienic conditions, packed under vacuum and cooked at fairly low temperatures. The cooking profile is not based on the need to destroy bacteria, but rather upon the culinary requirements of the product. This latter factor has meant that the shelf life of *sous-vide* products is normally limited to between one and three weeks.

Other technologies applied to minimally-processed foods include the addition of anti-microbial agents such as enzymes and other natural compounds that inhibit the growth of bacteria or prevent oxidation. Again, they constitute one step in an integrated hurdle approach that must follow the product from the farm to the consumer. This technology is gaining particular ground in Europe and will likely follow suit in North America. Unfortunately, this desire to employ a combination of less severe treatments in the hope that, together, they will ultimately accomplish the task of inhibiting undesirable organisms may lead to other problems.

Trends in Trade

The world today is characterized by intensive traffic of people and goods. Modern air, sea and ground transportation bring the most distant places of the earth in close contact with one another. This situation is desirable and of great benefit to us all. However, it also creates some dangers which did not previously exist. Among these is the movement of pests such as insects from country to country. To prevent or minimize this risk, countries have established quarantine measures supported by laws and regulations. This can create significant barriers to international trade and the free flow of plants and plant products.

Quarantine Measures

Because of the disastrous potential consequences of such pests, quarantine measures are enforced by law. Quarantine regulations prohibit the entrance of plants or plant products which might hide the unwanted pest from countries where it is known to exist. Inspections are carried out in the ports of entry with the objective of intercepting and destroying contaminated material. Exemption from such quarantine is granted only on assurance that agreed measures to disinfect the plant material have taken place. Such measures could consist of specific treatment of the material at the point of origin, in transit or at the port of entry.

Quarantine treatments generally fall into three categories.

1. Chemical treatments - these are fumigants for pests which occur inside or on the surface of the plant or plant product.

Until a few years ago, Ethylene-dibromide (EDB) was the most common example of this. In 1984, the U.S. banned its use on grain, after it was recognized as a possible carcinogen. Since then its use as an agricultural pesticide has been banned in most countries. Methyl bromide, another widely-used fumigant, has been categorized as an ozone depleter and is currently being phased out.

Non-chemical methods comprise the next two categories.

2. Physical treatments - Temperatures treatments applied in various ways.

Cold treatment as a quarantine measure for fruit flies is a rather old method which requires long-term storage of 10 to 15 days of the packed fruit (citrus) at temperatures ranging from 0.5°C to 2.0°C. Unfortunately, the method is three to five times more expensive than EDB and this long period may fail due to the breakdown of cooling or monitoring equipment. It is also relatively damaging to certain varieties of fresh fruits or vegetables which cannot tolerate excessive cold, such as papayas, avocados and mangoes.

High temperature treatments, such as double hot water dipping, are also used for mangoes and papayas, but fruit quality can suffer.

C. Ionizing radiation - Numerous investigations on the use of ionizing radiation for the disinfestation of fresh plant materials for quarantine purposes have been completed. Recent studies show that rather low dosages of ionizing radiation will control fruit fly problems. This makes the use of irradiation for quarantine treatment an excellent practical possibility.

The role of irradiation as a quarantine treatment of fresh fruits and vegetables was first evaluated internationally by a group of experts convened by FAO and IAEA in 1970 (IAEA, 1971). Although it was recognized at that time that irradiation was an effective quarantine treatment of such commodities, there was no economic incentive in using it commercially in view of the wide application of fumigation, especially using ethylene dibromide (EDB), to overcome quarantine barriers.

Following the ban of EDB in the USA in 1984, the International Consultative Group on Food Irradiation (ICGFI) convened a Task Force on Irradiation as a Quarantine Treatment, which eventually recommended a generic dose of 0.15 kGy (minimum) as a quarantine treatment of fresh fruits and vegetables against fruit flies of the Trephritidae family, and 0.30 kGy (minimum) against other arthropod pests, including mango seed weevils (ICGFI, 1986).

The effectiveness of irradiation as a broad spectrum quarantine treatment of fresh fruits and vegetables was recognized by the North American Plant Protection Organization (NAPPO) in 1989. In addition to NAPPO, other regional plant protection organizations which operate within the framework of the International Plant Protection Convention have endorsed irradiation as a quarantine treatment of fresh horticultural products at the Technical Consultation of Regional Plant Protection Organizations, held in San Salvador in 1992 (FAO, 1992).

The USDA issued a policy statement permitting irradiation to be used as a quarantine treatment of fresh fruits against major species of fruit flies, at specified minimum doses, regardless of host commodities and countries of origin (USDA, 1996). In response to the petition from Hawaii, the USDA approved the use of irradiation as a quarantine treatment of fresh papaya, lychees and carambola against fruit fly infestation in July 1997 (USDA, 1997). NAPPO also issued a standard on irradiation as a phytosanitary treatment of fresh horticultural produce in April 1997 (NAPPO, 1997).

Irradiation offers a broad scope as a quarantine treatment of fresh horticultural produce. The interest in using irradiation as a quarantine treatment of fresh and stored food products has increased recently because methyl bromide (MB), the most widely used fumigant to control insects in food and agricultural commodities, is being phased-out globally under the international treaty for the regulation of ozone depleting substances (Montreal protocol). Research data have demonstrated that irradiation could be used as a quarantine treatment instead of MB for a number of food and agricultural commodities.

There is some concern, however, at the new Codex Guidelines promulgated in July 1999, on the production of Organic Foods. These do not permit the use of irradiation. It is a subject that will hopefully be reconsidered in future.

On the other hand, there is good news in the lifting of the 10-year ban on irradiation in New Zealand and Australia.

There is no doubt that food irradiation has a strong future. Insect quarantine measures serve as only one example. There are several other applications for food irradiation in the reduction of post-harvest losses. Some other examples of the use of radiation to enhance the safety and quality of food are explained below:

Control of Sprouting and Germination

Control of sprouting is critical for the effective storage of certain commodities. Currently, sprout inhibition is carried out through a combination of chemical treatment and cold storage. Most people do not realise that many common foods such as potatoes receive chemical treatment to prevent sprouting. This is understandable because potatoes do not look like they have been treated and they are not labeled. Several chemical compounds are permitted for use as sprout inhibitors. Maleic hydrazide is applied in the field and works systemically. It is first absorbed by the leaves and then is trans-located to the tubers. Isopropyl-N-Chlorophenyl Carbamate or Chlorpropham is another chemical used, but is directly applied to potatoes before storage. Tetrachloronitrobenzene or Tecnazene which has come under fire from U.K. environmentalists, is also commonly used as a sprout inhibitor.

An alternative is ionizing radiation. It brings out a rather interesting issue. Many people who are concerned with the environment have a natural aversion to anything remotely related to nuclear energy. Yet, it seems logical that the irradiation of potatoes would be far more benign to the environment than is the current application of chemicals such as chlorinated hydrocarbons.

Controlling the germination of barley during malting is of considerable economic importance. Low doses applied to air-dried barley do not prevent the emergence of shoot tips and tendrils during malting, but do retard root growth. In this way, high quality malt can be obtained while the losses resulting from root growth are reduced. Since this effect of radiation processing persists for many months, treatment can be applied before the barley is put into storage, with the added benefit of destroying any insect pests that may be present in the grain. In fact, small radiation doses (0.01-0.10

kGy) stimulate the germination of barley, a result that can be used to shorten the malting process and increase the production capacity of malting plants.

Shelf-life Extension of Perishable Foods

One of the principal uses of food irradiation is for killing the microorganisms that cause spoilage or deterioration of the product. The shelf-life of many fruits and vegetables, meat, poultry, and fish and other seafood can be considerably prolonged by treatment with combinations of refrigeration and relatively low doses of radiation that do not alter flavor or texture. Many food spoilage microorganisms are killed at doses of less than 5 kGy. Various fresh fruits, including strawberries, mangoes and papayas, have been irradiated and marketed successfully.

Delaying Ripening and Aging of Fruits and Vegetables

Exposure to a low dose of radiation delays the ripening and/or senescence of some fruits and vegetables, thereby extending their shelf-life. The magnitude and even the direction of such changes depend on the size of the dose and the state of ripeness at the time of treatment. A measurable extension of shelf-life may be obtained with doses of 0.3-1.0 kGy. This level of exposure will increase the shelf-life of mangoes by about one week and that of bananas by up to two weeks. Maturation of mushrooms and asparagus after harvesting can be retarded with doses in the range of 1.0-1.5 kGy. Currently, chemical treatment is used to delay ripening.

Destruction of Parasites

Irradiation inactivates certain parasitic organisms that are responsible for both human and animal diseases. The parasitic roundworm *Trichinella spiralis*, which causes trichinosis and is found in pork, is inactivated by irradiation at a minimum dose of 0.15 kGy. Other parasites, including pork and beef tapeworms, the protozoan in pork responsible for toxoplasmosis, and various flukes that infest fish are rendered non-infective by low dose irradiation treatment.

Irradiation is beneficial in controlling the microbial contamination of dry food ingredients, and this improves the safety and storage properties of foods prepared with them. Spices, dry vegetable seasonings, herbs, starch, protein concentrates, and commercial enzyme preparations used in the food industry are very often heavily contaminated with spoilage and pathogenic organisms.

Summary

Under the present circumstances, irradiation will find practical uses only in advanced post-harvest systems where the distribution infrastructures are reliable and highly efficient. With the possible exception of plantation crops such as large scale roots and tubers, coffee, tea, and spices or seafood, food irradiation has very limited use at present in developing countries. The focus in these countries will continue to remain on improving of small-scale village storage and marketing systems to ensure minimum losses.

The benefits that flow from the establishment of an orderly and efficient post-harvest system are very significant. Such a system creates the climate of confidence required to persuade subsistence farmers to produce beyond what they require for themselves. It substantially increases the availability of food, it improves people's well-being through improved nutrition, and it reduces the outflow of foreign currency, resulting in general economic benefits for all.

Rationally conceived, well-ministered post-harvest systems have helped to make the supply of food crops more consistent and reliable, as well as discouraging cycles of boom and bust, with resultant wildly fluctuating price levels. In addition, an orderly post-harvest system offers many opportunities for increased rural and urban employment, that is, employment in harvesting, grading, storing, transportation, processing and marketing.

A good post-harvest system has other benefits too. It reduces spoilage and will enhance the acceptability, utility and nutritional quality of foods. It leads to the establishment of entirely new industries that produce processed foods and provides a consistent year-round demand for indigenous crops.

In summary, the field of post-harvest technology applied to horticultural crops remains very much alive - whether we consider the latest new and emerging technologies such as food irradiation and high pressure processing or simply methods of improved on-farm storage, using structures of modern design or even farm-level solar-refrigerated storage. This fact equally applies to treatments which extend shelf life using various chemical or natural compounds. The challenges of free trade will require smaller commercial operations to seek strategic new linkages in order to survive. One thing we can say for certain - this is a sector which will continue to meet the challenges of ensuring a sustainable and nourishing future for us all.

REFERENCES

[1] FAO. 1992. Report of Fourth Technical Consultation Among Regional Plant Protection Organizations, San Salvador, El Salvador, 11-15 May 1992. AGPP/1992/M/3. FAO, Rome.
[2] IAEA, 1971. Disinfestation of Fruit by Irradiation. Proceedings of a Panel of Experts, Honolulu, Hawaii, 1970. IAEA, Vienna.
[3] ICGFI. 1986. Report of the Task Force Meeting on Irradiation as a Quarantine Treatment, convened by the International Consultative Group on Food Irradiation (ICGFI), Chiang Mai, Thailand. IAEA, Vienna.
[4] ICGFI. 1991. Report of the ICGFI Task Force on Irradiation as a Quarantine Treatment of Fresh Fruits and Vegetables, Bethesda, Maryland. ICGFI Document 13. IAEA, Vienna.
[5] NAPPO, 1997. NAPPO Standards for Phytosanitary Measures: Guidelines for the use of irradiation as a phytosanitary treatment. NAPPO Newsletter, Vol. 17, No. 3 (July 1997).
[6] USDA 1995. The Application of Irradiation to phytosanitary Problems: Position Discussion Document IV. USDA Animal Plant Health Inspection Service, Washington, D.C., September 1995.

[7] USDA, 1996. The Application of Irradiation to Phytosanitary Problems: Notice of Policy. 7 CFR Part 319, Federal Register Vol. 61, No. 95, pp 24433-24439 (May 15, 1996).

Low Energy Electron Irradiation of Food for Microbial Control

T. HAYASHI and S. TODORIKI-SUZUKI
National Food Research Institute
Maff, Japan

ABSTRACT

"Soft-electron", new term we have created means an electron with an energy of 300 keV or lower. Soft-electrons with extremely small penetration capacity will open a door to novel sanitary and phytosanitary treatments for the first century of next millennium. Use of soft-electrons can supplement food irradiation technology developed and commercialized in the last decades of this century. Irradiation can decontaminate spices, herbs, dehydrated vegetables with little adverse effects on flavor and color, and blood plasma and whey protein with little functional change. Radiation treatment, however, still has a few drawbacks; degradation of starch and oxidation of lipids, so irradiation can not apply to decontamination of grains with high starch contents and seeds (e.g. soy bean and sesame) with high lipid contents. Irradiation adversely affects germination of seeds, as well.

Soft-electrons can eradicate microorganisms on the surface of dry food ingredients such as grains, dehydrated vegetables, spices, pulses and tea leaves with little quality deterioration. The penetration depths of the soft-electrons are too small to cause detrimental effects on the ingredients. Treatment of seeds with soft-electrons results in extremely low levels of microbial contamination and the same extents of germination and growth, as compared with untreated seeds. Soft-electrons at lower dose inactivate eggs, larvae, pupae and adults of insects such as red flour beetle and Indian meal moth. Adults of adzuki bean weevil do not emerge from the beans which have been infested with the eggs and exposed to soft-electrons, while a lot of adults emerge from untreated beans infested with the eggs. Soft-electrons can inhibit sprouting of potatoes.

The portion of grains exposed to electrons are removed as **husk and bran** by dehusking and/or milling. Potatoes are usually peeled and then used for processing or cooking. Therefore, chemical compounds formed in food by electron exposure are removed by dehusking, milling or peeling, and no such compound resides in the edible parts, which suggests the safety of foods treated with soft-electrons.

Electrons with such low energies do not require a thick shield due to their low penetration capacity, which enables in-line decontamination at food processing plants. The total cost for construction of a facility with an accelerator and a shield is less than US$ one million, much cheaper than ordinary irradiation facilities with gamma-rays or high-energy electrons. The costs of soft-electron treatment are estimated at US$50-200/ton and US$10/ton for decontamination and sprout inhibition, respectively.

INTRODUCTION

Grains, pulses, spices and dehydrated vegetables are frequently contaminated with heat-resistant bacterial spores, although they are widely used for food processing. Bacterial spores contaminating these dry food ingredients can not be inactivated completely by heating processes such as steaming and boiling at food processing plants. Thus dry ingredients for food processing should be decontaminated to prevent food spoilage and food-borne diseases. Irradiation with gamma-rays or electron beams can effectively inactivate bacterial spores contaminating spices and dehydrated vegetables with minimal quality alteration, and irradiation of spices is practiced on a commercial basis in more than 20 countries [1)]. However, irradiation of food products other than potato is not approved in Japan. Irradiation at high doses for decontamination is not applicable to grains such as rice and wheat. Ionizing radiation causes oxidation and degradation of components such as lipid and starch molecules[1)-3)], which results in quality deterioration of grain products such as noodles and rice cake in terms of flavor and texture. Treatment with super-heated steam is the only method widely used for decontaminating dry ingredients in Japan. However, super-heated steam sometimes brings about changes of flavor and color of dehydrated vegetables and herbs. No method has been developed for decontaminating all the dry food ingredients.

Since most of the microorganisms contaminating dry food ingredients reside on their surfaces, the inner parts do not have to be exposed to heat, gas or radiation for decontamination. Low energy electrons with limited penetration capacity can reach the surface layer of food but not their inner part. It is, therefore, expected that low energy electrons can decontaminate the food ingredient with little quality deterioration.

In this chapter, we will review our studies on the low energy electron irradiation as a means of decontamination. In the first section we will describe the basic concept of the low energy electron irradiation, including equipment for irradiation. The second section deal with the efficacy of low energy electron treatment on decontaminating dry food ingredients, especially sterility and quality of rice treated with low energy of electrons. The final section refers to the application of this method to the disinfection of seeds for sprouts production, which is recently recognized as a source of food-born diseases.

Concept of low energy electron (soft-electron) irradiation

The penetration capacity of an electron beam is controlled by the energy; electrons with lower energy display lower penetration capacities [4)]. We have defined low-energy electrons at 300 keV or lower as "soft-electrons". Soft-electrons reach only the surface of food, and the quality changes of food caused by such electrons are expected to be much more limited than those caused by gamma-rays or high-energy electron beams with much higher penetration capacities.

Equipment for low energy treatment

Dry ingredient samples were treated with soft-electrons under rotation to expose all the sample surfaces to electrons with low penetration capacities. A grain rotator was developed which enabled samples to rotate by shaking and vibrating them simultaneously at variable speeds (Fig. 1). The rotator was placed under the window of a Van de Graaff electron accelerator (Nissin High Voltage Engineering Co., Ltd., Kyoto, Japan), which generated electrons at acceleration voltages of 170-300 kV. The distance between the window of the electron accelerator and the plastic tray of the grain rotator was 17 cm.

Model indicator ball was made from paraffin and radiation sensitive dye (methyl-yellow), in order to visualize the absorption of electron energy by the color change (yellow to red). When the balls were treated with electrons at different acceleration voltages on the tray of the grain rotator, the surface color of the balls uniformly changed at any voltages, which indicates that the surface of the balls were uniformly exposed to electrons. When the balls were sliced into the sections, the inside part remained yellow, indicating that low energy electrons can reach only the limited (outside) part of the balls. The penetration capacity was depending on the energy of the electrons. Electrons at acceleration voltage of 200 keV only reached to a few hundredth mm from the surface, while electrons at 1500 keV can reach to the center of balls (Figure 2a). Similarly, the distribution and penetration of electrons within the rice grain were obviously shown by the color change of the section of the rice grain stained with methyl yellow. (Fig. 2 b)

Penetration Capacity of Low energy electron

Penetration capacities of soft-electrons at different energies were determined based on depth-dose curves. Several pieces of radiochromic film dosimeter (RCF)(5.94 mg/cm2, FWT-60-00, Far West Technology Inc., Goleta, California, USA) were stacked together in layers at the bottom of the plastic tray of the grain rotator (C of Fig. 1) which was placed under the window of the electron accelerator at a distance of 17 cm, and irradiated with electrons for 60 min under various conditions (170 kV, 4 mA; 180 kV, 8 mA; 190 kV, 10 mA; 200 kV, 14 mA). Absorbances at 510 nm of all RCF films before and 30 min after irradiation were measured and the dose absorbed by each RCF film was determined according to the method of McLaughlin et al.[5].

Energies of electrons at a distance of 17 cm (air) from the window (50 mm thick titanium) of the electron accelerator were lower than those of the electrons at the window (acceleration voltage). The energies of electrons irradiating samples at 17 cm from the window were estimated to be 60, 75, 90, 100, 130, 160 and 210 keV for acceleration voltages of 170, 180, 190, 200, 225, 250 and 300 kV, respectively, based on the mass stopping power of air and titanium[6),7)]. Depth-dose curves of electrons at various energies were developed by plotting all the absorbed doses determined with RCF films (Fig. 3) [8]. The penetration capacity of electrons at 60 keV was about 6 mg/cm^2 and that at 75 keV was about 10 mg/cm^2, while those at 90 and 100 keV were lower than 17.82 mg/cm^2 (5.94 mg/cm2 x 3 pieces). Doses absorbed by the first RCF film for 1 h were about 30, 58, 70 and 110 kGy at 60, 75, 90 and 100 keV, respectively (Fig. 3)[8].

Decontamination of dry food ingredients with low energy electrons

Sterility and viscosity of grains exposed to soft-electrons

Energies of electrons necessary to reduce microorganism levels to less than 10 CFU/g were 75 keV for brown rice, 160 keV for rough rice, 75 keV for wheat and 160 keV for buckwheat (Table 1)[9]. The results suggested that most of the contaminating microorganisms resided in the region that the electrons with such low energies could reach. Gamma-rays at 7.5-12.5 kGy were necessary to achieve the same levels of sterility.

Viscosity of heat-gelatinized grain suspensions decreased with the energy of electrons (Table 2)[9]. Viscosity of brown rice and wheat treated with electrons at 75 keV was almost the same as that of untreated samples. Viscosity of rough rice and buckwheat exposed to electrons at 130 keV was slightly lower than that of untreated samples, but much higher than that of the samples irradiated with gamma-rays at 10 kGy. The viscosity values of a grain suspension which was heat-gelatinized under an alkaline condition is a parameter for starch degradation [10, 11]. The results suggested that electrons with minimum energy for decontamination did not degrade starch molecules inside the grains.

The applicability of soft-electrons to wheat depended upon the variety of wheat. Soft-electrons could decontaminate wheat grains of Australian Standard White (ASW), Shirogane and Western White (WW) varieties, but could not decontaminate those of Dark Northern Spring (DNS), Norin No. 61 and No. 1 Canadian White (1CW) varieties (Table 3). No relationship was observed between the application of soft-electrons and the sensitivity of the contaminating microorganisms to gamma-rays. The difference in the application of soft-electrons was ascribed to the structure of the wheat grain. In varieties such as DNS, Norin No. 61 and 1CW, microorganisms would reside in areas that soft-electrons did not reach.

Quallity of milled rice prepared from brown rice treated with soft electrons

Both Koshihikari and Nihonbare could be decontaminated with soft-electrons even at 60 keV [8]. No significant difference in the viscosity was observed at any milling yield between the control and the rice grains exposed to electrons at 60-90 keV. Milling reduced the effect of electrons on the viscosity; milling of rice grains at a yield of 88% did not result in a significant difference in the viscosity between untreated samples and 100 keV-electron treated samples [8]. However, milling did not affect the viscosity of rice irradiated with gamma-rays. The results suggested that soft-electrons degraded starch molecules near the surface of rice grains, which could be removed easily by milling. On the contrary, gamma-rays degraded all the starch molecules in rice grains most of which could not be removed by milling[8].

Thiobarbituric acid (TBA) value is a parameter of lipid oxidation. TBA value of brown rice increased with the energy of electrons. TBA values of brown rice samples exposed to electrons at 60-100 keV were significantly higher than that of untreated control (Table 4)[8]. The use of gamma-rays at 7.5 kGy resulted in a higher TBA value than that of electrons with energies of 60-100 keV. Milling decreased the TBA values of all the samples, especially those of electron-treated samples. Accordingly, the differences in the TBA values between the control and the electron-treated rice grains decreased markedly after milling. The difference in the TBA values between the control and the rice samples exposed to electrons at 60 keV was not significant at a milling yield of 90% or lower, and the difference between the values of the control and the rice samples exposed to electrons at 75 keV was not significant at a milling yield of 88%. The results suggested that most of the lipids oxidized by electrons at 60 and 75 keV were removed by milling at yields of 90 and 88%, respectively.

Hardness and stickiness under low and high compressions of cooked rice grains (90% milling yield) exposed to electrons at 60-75 keV were almost the same as those of the control (Table 5) [8]. Hardness and stickiness under low and high compressions of gamma-irradiated samples were lower than those of the control. Hardness and stickiness under low compression are parameters for rheological properties of the surface of cooked rice grains, and those under high compression are parameters for the properties of overall cooked rice grains [12]. The results showed that rice grains which were exposed to electrons at 60-75 keV and milled at a yield of 90% displayed the same rheological properties as the control. The results shown in Tables 4 and 5 indicate that milling at yields of 88-90% removed the portion of rice exposed to electrons at 60 keV which could eradicate most of the microorganisms contaminating brown rice.

Decontamination of other food ingredient

Soft-electrons at 100 keV reduced the microbial loads of shredded dehydrated vegetables to levels lower than 10 CFU/g, although the time necessary for electron treatment was different (Table 6). Longer duration of electron treatment was necessary for dehydrated vegetable samples which required a higher dose of gamma-rays for disinfection. The microorganisms contaminating dehydrated vegetables showed the same resistance to soft-electrons as to gamma-rays. Black pepper could be decontaminated with electrons at 210 keV, while white pepper, basil and coriander could be decontaminated at 100 keV. Pulses such as soybean, adzuki bean and black soybean could be decontaminated with electrons at 60 keV.

Decontamination of seeds with low energy electrons

Recently sprouts have emerged as a recognized source of food-borne disease. The outbreaks of *Salmonella* and *E. coli* O157:H7 infection were associated with eating raw sprouts, such as alfalfa and radish sprouts. It is believed that the seeds from which sprout are derived are often the source [13,14]. It is, therefore, urgent subject to develop decontamination method of sprout seeds without compromising seed's germination potential. Moreover, the establishment of the method for decontamination of plant pathogen in crop seeds would contribute to the increase in agricultural production.

Seed of mung bean could be decontaminated with electrons at 75 keV. And the percentages of sprouting seeds were not affected by electrons at 100keV (Table 7), and the volume of the sprouts of mung bean after incubation for 4 days at 30 °C were almost the same as those of the controls (untreated samples) (Figure 4). The results suggest that soft electrons necessary for decontamination did not adversely affected the sprouting of seeds or growth of sprouts.

Similarly, electrons at 60-75 keV reduced total microbial counts of alfalfa and radish sprouts (kaiware-daikon) seed to undetectable levels without any detrimental effect on the germination ability.

In addition to the decontamination of sprout seeds, treatment of crop seeds with soft-electron is applicable to the reduction in the number of phytopathogenic organisms. Treatment of rice seeds with 100kV electron reduced the outbreak of blast disease and damage of nematode without any damage to the growth and harvest of rice crop. The result indicates the possibility of soft-electron treatment as a means of disinfecting seeds alternative to chemicals.

Conclusion

Low energy electrons (Soft-electrons) eradicate microorganisms on the surface of dry food ingredients such as grains, dehydrated vegetables, spices and pulses with little quality deterioration. The penetration depths of the soft-electrons are too shallow to cause detrimental effects on the ingredients when the sizes of the ingredients are taken into consideration. Soft-electron treatment is one of the efficacious techniques for decontaminating dry food ingredients, except for powdered products. The portions of grains exposed to electrons are removed as husk and bran by and/or milling. Therefore chemical compounds formed by electron exposure are removed by dehusking or milling and no such compounds occur in the edible parts of grains, which suggest the safety of foods treated with soft electrons. In addition to decontamination of dry food ingredient, soft-electrons can disinfect the seeds without deteriorating their germination and growing ability.

Electrons with such low energies do not require a thick shield due to their low penetration capacity, which enables in-line decontamination at food processing plants. Facility to expose all the ingredient surfaces uniformly to soft-electrons on a large scale in a short period of time should be developed to industrialize this decontamination technique.

References

1) WHO 1994. Safety and nutritional adequacy of irradiated food. Geneva, Switzerland : World Health Organization.
2) Diehl, J. F. 1990. Safety of irradiated foods. New York, USA: Marcel Dekker, Inc.
3) Urbain, W. M. 1986. Food irradiation. Orlando, USA: Academic Press, Inc.
4) Hayashi, T. 1991. Food irradiation. London, UK: Elsevier Science Publisher, Inc., pp.167- 216.
5) McLaughlin, W. L. et al. 1979. "The response of plastic dosimeters to gamma rays and electrons at high absorbed dose rates." Radiat. Phys. Chem., 14: 535-550.
6) ICRU. 1984. ICRU Rep. 37. Bethesda, USA : International Commission on Radiation Units and Measurements.
7) Tabata, T. and Tanaka, R. 1990. Radiation dosimetry of electron beams for radiation processing. ed. Moriuchi.Y. Tokyo, Japan: Chijinshokan Publisher, Inc., pp109-135 [In Japanese].

8) Hayashi, T. et al. 1998. "Rheological properties and lipid oxidation of rice decontaminated with low-energy electrons." J. Food Protect. 61(1): 73-77.
9) Hayashi, T., Takahashi, Y. and Todoriki, S. 1997. "Low-energy electron effects on the sterility and viscosity of grains." J. Food Sci., 62: 858-860.
10) Hayashi, T., Todoriki, S. and Kohyama, K. 1993. "Applicability of viscosity measuring method to the detection of irradiated spices." J. Jpn. Soc. Food Sci. Technol., 40: 456-460.
11) Hayashi, T., and Todoriki, S. 1996. " Detection of irradiated peppers by viscosity measurement at extremely high pH." Radiat. Phys. Chem., 48: 104-104.
12) Okadome, H., Toyoshima, H. and Ohtsubo, K. 1996. "Many-sided evaluation of physical properties of cooked rice grains with a single apparatus." J. Jpn. Soc. Food Sci. Technol., 43: 1004-1011 [In Japanese with English summary].
13) Urtzweil, P., 1999. "Questions keep sprouting about sprouts." *FDA Consum Jan-Feb*, 33(1) :18-22.
14) Mahon, B.E., Ponka, A., Hall, W.N., Komatsu, K., Dietrich, S. E., Aiitonen,A., Cage, G., Hayes, P.S. Lambert-Fair,M.A., Bean, N. H., Griffin, P.M. and Slutsker, L. 1997. "An international outbreak of Salmonella infection caused by alfalfa sprouts grown from contaminated seeds." *J. Infect. Dis*. Apr.1 175 (4): 876-882.

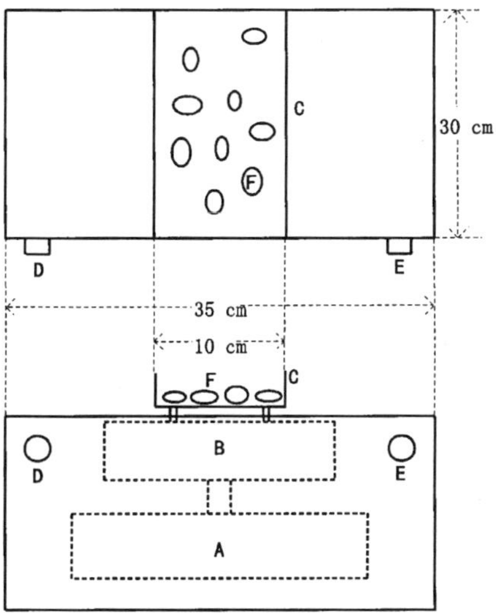

Fig. 1. Grain rotator for treating grains with low-energy electrons. (Hayashi et al. 1998)[8]
A ; Shaker, B ; Vibrator, C ; Plastic tray, D ; Speed controller for shaker,
E ; Speed controller for vibrator, F ; Grain.

a

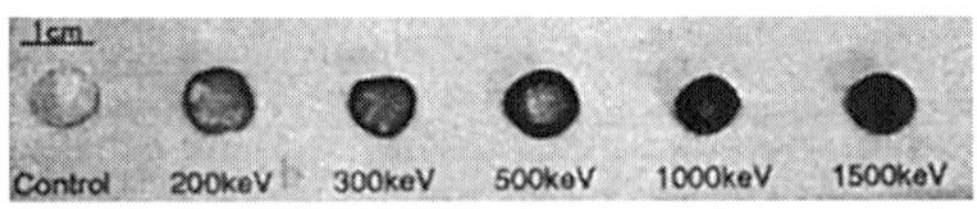

b

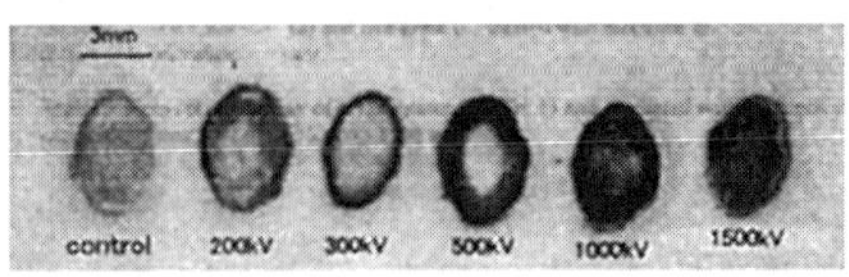

Fig. 2. Section of indicator ball (a) and rice grain (b) treated with electrons at different acceleration voltages.

Samples were set on the tray of grain rotator (see Fig.1) and irradiated with electrons at acceleration voltages of 200 keV to 1500 keV.

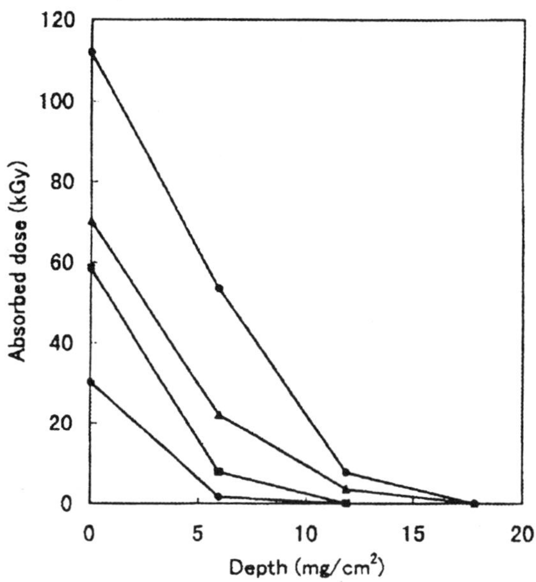

Fig. 3. Depth-dose curves of electrons determined with radiochromic film dosimeter (RCF). (Hayashi et al. 1998)[8)]
u ; 60 keV, 4 mA, n ; 75 keV, 8 mA, s ; 90 keV, 10 mA, l ; 100 keV, 14 mA.

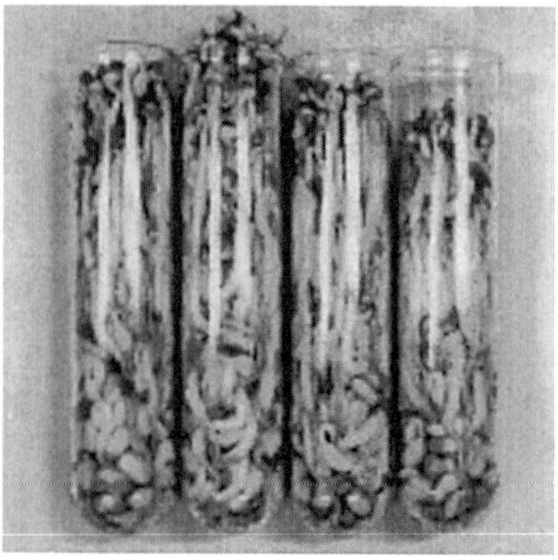

Fig.4. Sprouts of mung bean after incubation for 4 days at 30°C.

Economics of Food Irradiation

P. KUNSTADT
MDS Nordion
Kanata, Ontario, Canada

ABSTRACT

The number of products being radiation processed worldwide is constantly increasing and today includes such diverse items as medical disposables, spices, meats, seafood and potentially, fruits and vegetables and waste products. This range of products being processed has resulted in a wide range of irradiator designs and capital and operating cost requirements. This paper discusses the economics of food irradiation applications and the effects of various parameters on unit processing costs. It provides a model for calculating specific unit processing costs by correlating known capital costs with annual operating costs and annual throughputs. It is intended to provide the reader with a general knowledge of how unit processing costs are derived, and enough information to assist in the choice of the irradiator model best suited for specific needs.

1. INTRODUCTION

Food irradiation is now well established along with other methods of food preservation, opening up opportunities for commercial application. Today, there is enough evidence that gamma irradiation can successfully reduce spoilage, disinfest, sanitize and extend shelf-life of foods. Having established the technical feasibility of this process, one of the biggest questions a potential user would have is that of economic feasibility. This paper discusses various processing parameters and their effect on the economics of food irradiation, using as a model scenario the low dose irradiation of food for sprout inhibition and fruit disinfestation.

The number of radiation processed food products today includes items such as fruits and vegetables, bulk and retail spices, meats, seafood, grains and cereals. The products differ in their radiation dose requirements, densities, dose maximum to minimum ratios, sensitivities to handling, temperature tolerances, as well as specific packing configurations. These variations create many different requirements in the irradiator design. Obviously, just as with any other type of machinery, irradiators are most efficient when they are specialized, i.e. they are most efficient when designed to handle a limited product density range at an established dose, at a steady throughput rate.

2. FOOD IRRADIATION PARAMETERS

The parameters associated with food irradiation will have an effect on the irradiator design and, therefore, on the economics of food irradiation.

(1) *Applied doses* vary significantly according to the effect required. For example, sprout inhibition doses are usually below 0.15 kGy (15 krad), whereas disinfestation doses range

up to 0.75 kGy (75 krad) and spice sanitization doses sometimes surpass 10 kGy (1 Mrad). Often, doses for the same food vary according to its geographic origin.

(2) *Packing densities* of foods are relatively high when compared to medical disposable products. This presents new challenges to equipment manufacturers in structural design and radiation physics.

(3) *Handling and distribution* conditions of foods will often dictate the physical form in which the product will be irradiated. Many products are perishable and require a short residence or hold-up time at the irradiation facility. Many are susceptible to physical damage and therefore irradiator designs need to minimize the effect of product handling. Preferably, the irradiator should accommodate the products without repackaging or restacking.

(4) *Dose uniformity*: Many food products have a well defined maximum tolerance to radiation. In many cases, there are legislative restrictions on maximum doses. The equipment must ensure a good dose uniformity; i.e. it must be able to deliver the minimum effective dose while, at the same time, not exceeding regulatory limits or product tolerance doses.

3. IRRADIATION EQUIPMENT

The preceding parameters along with unique customer information translate into irradiator design requirements. For simplicity, we have selected the following typical examples of basic irradiator designs, which are generally suitable for food irradiation:

(1) Tote box concept - Figure 1
(2) Carrier concept - Figure 2
(3) Pallet carrier concept - Figure 3
(4) Pallet conveyor concept - Figure 4
(5) Horizontal source concept - Figure 5

In the tote box irradiator concept, the product is packed in metal or fibreglass boxes called totes. A tote has typically a volume of approximately 0.75 m^3 and can accommodate a multitude of product boxes. Actual sizes vary to accommodate users' specific needs. In the batch version, the totes are transported into the irradiation chamber on handcarts and placed on the irradiator mechanism manually. In the automatic mode, the totes are transported in and out of the irradiation chamber on automatic conveyors. In both cases the irradiator mechanism indexes the totes around the radiation source in a way that ensures maximum utilization of the source and best dose uniformity. Tote box irradiators are designed with `overlapping product' source configuration. This configuration offers superior cobalt utilization efficiency. It requires the product totes to pass around the radiation source at least four times: once at the upper and once at the lower level on both sides of the source. Figure 1 illustrates the automatic version of the tote box irradiator concept with a double pass system, where the product totes pass around the radiation source eight times.

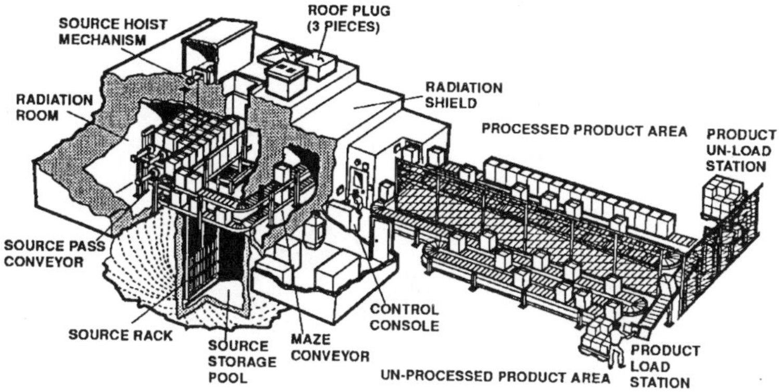

FIG 1. Tote Box Irradiator - Automatic

The carrier irradiator concept uses tall (approximately 3 m high) aluminum carriers. The carriers are suspended and move through the irradiation chamber and around the cobalt source on an overhead monorail system. The carriers have a large internal volume and can thus accommodate a wide range of product sizes. In the batch version of this concept, the cariers are moved in and out of the irradiation chamber manually; in the automatic version, the carriers are moved by pneumatic or hydraulic cylinders. Most carrier irradiators are designed with overlapping source configuration to allow maximum throughput. The automatic version of the carrier irradiator concept is illustrated in Figure 2.

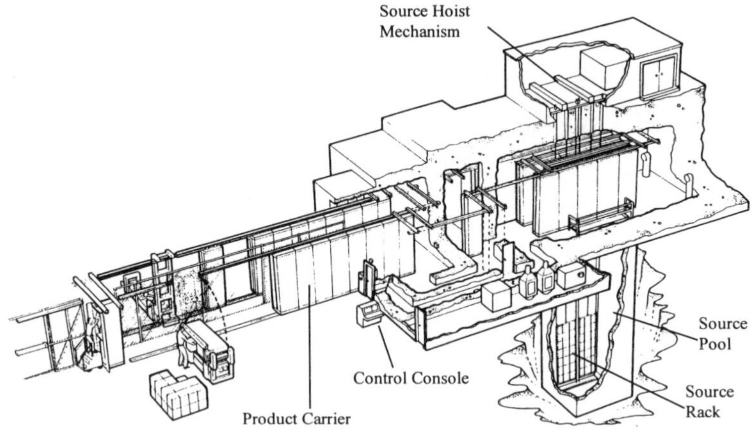

FIG. 2 Carrier irradiator - automatic, with vertical and horizontal load/unload positions

The pallet irradiator concept is similar to a carrier, but its carrier configuration makes it possible to irradiate product stacked on standard transportation pallets. This concept reduces the labour intensive handling of the product often required in tote and carrier irradiators. The loaded pallets are placed in suspended aluminum containers even larger than those used in the carrier concept, and conveyed through the irradiator on a monorail system and around the source in a specially indexed sequence, to compensate for the large carrier dimensions and to ensure good dose uniformity. The processing capability of the pallet irradiator per curie of Cobalt 60 is not as high as that for the tote box and carrier designs. However, it has the advantage of being the least labour intensive of the three concepts and it is capable of much greater throughputs. Various source and carrier geometry alternatives are available to address specific dose and throughput requirements. Cobalt utilization efficiencies vary from case to case. As a general rule, pallet carrier irradiators are designed with `overlapping source' configurations. However, there are designs aimed at improved cobalt utilization efficiency by using `overlapping product'. Switching stations must be added in this case to allow for each pallet to make its journey through the irradiation chamber in both the upper and the lower compartment of the carrier. An example of a pallet irradiator with overlapping source was chosen for this scenario. It is illustrated in Figure 3.

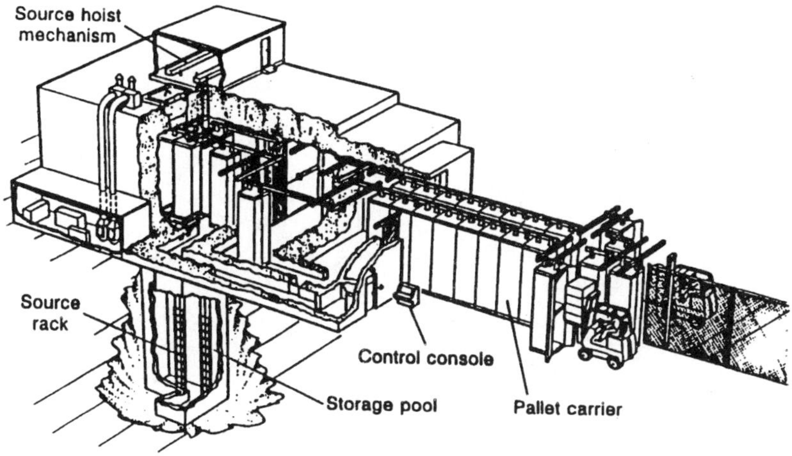

FIG. 3. Pallet irradiator — automatic.

The pallet conveyor system is an adaptation of the basic tote box irradiator. The conveyor is designed to handle product on standard transportation pallets, improving product handling efficiency while offering the high cobalt utilization inherent in the `overlapping product' concept. It combines the simplicity of operation of the pallet irradiator with the efficiency of the tote box irradiator. An example of a pallet conveyor irradiator is illustrated in Figure 4.

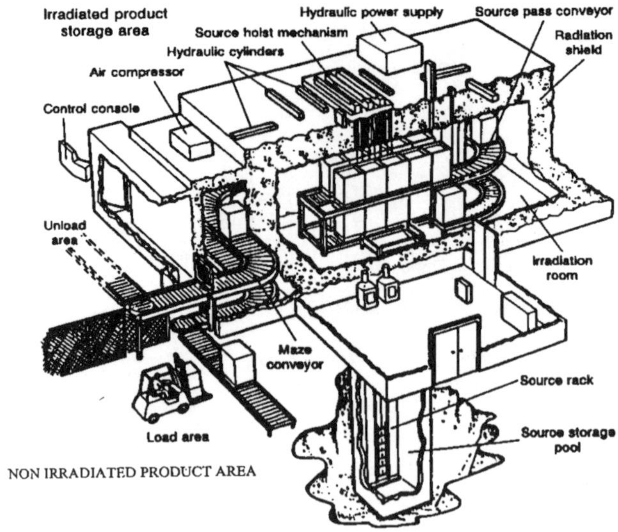

FIG. 4 Pallet Conveyor Irradiator

The automated versions of these irradiator concepts have the highest throughput capacity, although they have somewhat higher capital costs. The batch versions require that the irradiators be shut down for product loading and unloading, whereas the automated versions can utilize the source for up to 8000 hours annually. In calculations throughput this paper, we have assumed a full utilization of 8000 hours per annum, a utilization rate regularly achieved by commercial irradiator companies.

The horizontal source concept is specifically designed for most efficient radiation processing of meat and poultry products. It ensures the best dose uniformity ($\leq$1.5:1) for typical product configurations. The horizontal source rack allows products to be irradiated in their normal orientation, eliminating the need for flipping or turning boxes, thus preventing fresh meat products from slumping or becoming damaged. This conveying mechanism follows a continuous helical path through four levels on two sides, for a total of eight passes by the source pack. It is possible to adjust the position of the carriers to accommodate products heights from 8"-16" (20-41 cm). Figure 5 illustrates a typical horizontal source irradiator.

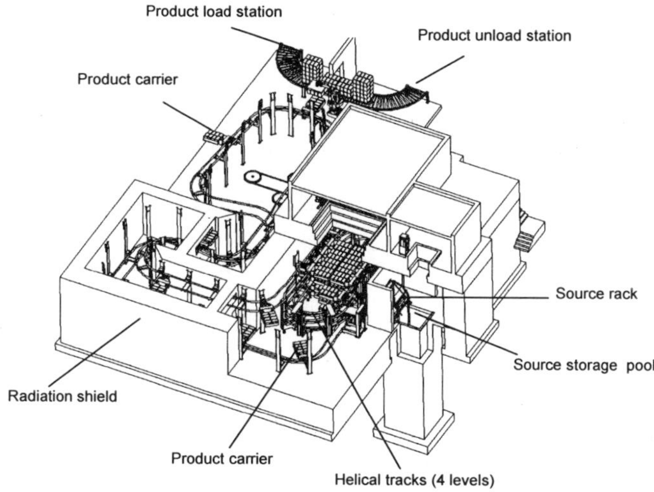

FIG. 5 HORIZONTAL SOURCE IRRADIATOR

4. COSTS

4.1 Capital Costs

Capital costs for a typical free standing food irradiation facility consist of the cost of hardware (irradiator, totes or carriers, conveyors, control systems and auxiliary equipment), Cobalt 60 (including container rental and transport), land (1.5 - 2.5 acres), radiation shield and warehouse.

For simplicity, costs of the land and warehouse have been excluded from the calculations, as they vary significantly from region to region.

4.2 Operating Costs

Operating costs include salaries, utilities, maintenance, taxes/insurance, interest, Cobalt 60 replenishment and miscellaneous operating expenses. In our calculations, we assumed salaries of US$85,000/annum for managers, US$65,000/annum for quality control and radiation safety officers, US$40,000/annum for operators and US$22,000/annum for handlers. The number of workers varies with equipment selection and throughput. Cobalt-60 replenishment is calculated at 12.3% per annum.

4.3 Total Processing Costs

The total processing cost equals cost of money, plus annual operating cost, plus depreciation of capital. Depreciation costs are calculated based on a 10 year amortization of equipment, 15 year amortization of cobalt and 25 year amortization of buildings. For simplicity, we have chosen linear amortization models. Cost of money is calculated as 10% of total capital. It represents either the cost of borrowing or the loss of interest income from the capital. Worksheet Nos. 1, 2,3, 4 and 5 provide examples of how the total processing cost is determined.

4.4 Unit Processing Costs

The unit processing cost equals the total annual processing cost divided by the annual throughput.

Examples of unit processing cost calculations for the typical tote box, carrier, pallet carrier and pallet conveyor irradiator designs, are shown in Worksheet Nos. 1, 2, 3, 4 and 5, respectively. For each specific scenario and product configuration, a separate worksheet applies. For example, in Worksheet 1a, we have chosen a packing density of 0.4 g/cm^3, which translates into cobalt utilization efficiency of 41%, for the typical tote box irradiator. We have chosen a low dose of 0.15 kGy (0.015 Mrad) for fruit disinfestation or sprout inhibition and a throughput of 60,000 tonnes/annum.

For this specific case, 51,311 Ci[2] of Cobalt 60 is required, for 8000 hr/annum operation. Cobalt source requirements (Ci) are calculated using a formula, taking into consideration the dose required (Mrad), throughput (kg/h) and cobalt utilization efficiency (%) :

$$S(\text{Ci}) = \frac{187 \times D(\text{Mrad}) \times T(\text{kg/h})}{\eta\ (\%)}$$

Tables I to V are based on numerous calculations performed as described above, for a variety of densities, doses and throughputs. Each unit processing cost figure in Table I, for example, is a result of a complete calculation, such as depicted in Worksheets 1a and 1b. Similar relationships exist for all the other sets of Worksheets and Tables. It should be noted that some simplifications have been made, for clarity. For instance, while there is no difficulty achieving the plant utilization of 8000 hr/annum, it is somewhat idealistic to assume that the source strength will precisely match the dose and throughput requirements at all times. Each irradiator type has its advantages and limitations. For example, the tote box irradiator is very flexible and efficient, but has a mechanical limit of 50-60 totes/hour, resulting in a maximum throughput of slightly below 120000 tons/annum at a density of 0.4 g/cm^3. Table I reflects this phenomenon. No such consideration is necessary in the case of pallet and carrier concepts. In most cases, the mechanical limits of the equipment exceed warehouse handling capacities.

---

[2] 1 Ci = 3.7 x 10^{10} Bq.

The pallet irradiator concept offers excellent throughout capacity and low labour requirements but present a challenge in terms of dose uniformity. They are ideal for processing fruits and vegetables or pallets, but require considerable product stack adjustments for treatment of meat products which generally require tighter dose uniformity as meat products are much more efficiently processed in the horizontal source irradiator models.

Worksheet No. 1a

ESTIMATE OF GAMMA PROCESSING COST FOR A TOTE BOX IRRADIATOR ($US)

Fruit & Vegetable Disinfestation Scenario

A. INPUT VARIABLES

1	Cobalt	$ 1.46	per Curie
2	Replenishment Cobalt	$ 1.46	per Curie
3	Irradiator Efficiency	41%	
4	Annual Operating Hours	8000	hours
5	Minimum Dose	0.15	kGy
6	Throughput	60,000	tonnes/annum
7	Container Rental Charge	$ 2,000.00	
8	Number of Product Handlers	16	
9	Cobalt Required	51,311	Curies
10	Replenishment Cobalt Required	6,311	Curies
11	Product Density	0.4	g/cc

B. CAPITAL ITEMS

12	Irradiator					$ 900,000
13	Tote Boxes	100 totes	@		550	$ 55,000
14	Auxiliary Equipment					$ 160,000
15	Warehouse Conveyors	400 feet	@		700	$ 280,000
16	Other					
17	Cobalt 60 Requirement (Ci)				51311 Curies	
18	Cobalt 60 Cost	51311 Ci	@	$	1.46	$ 74,914
19	Container Rental	1 cont.	@	$	2,000	$ 2,000
20	Land (1.5-2.5 acres)					
21	Radiation Shield					$ 800,000
22	Warehouse					
23	Shipping					$ 45,000
24	Start Up Cost					$ 40,000
25	**TOTAL CAPITAL**					**$ 2,356,914**

C. OPERATING ITEMS

26	Salary:	Manager					$ 85,000
		RSO					$ 65,000
		Operators	4	@	$	40,000.00	$ 160,000
		Product handlers	16	@	$	22,000.00	$ 352,000
27	Utilities and Communications						$ 15,000
28	Maintenance and Repairs						$ 60,000
29	Taxes and Insurance						$ 20,000
30	Operational Supplies						$ 30,000
31	Administration						$ 30,000
32	Cobalt 60 Replenishment		6311 Ci	@	$	1.46	$ 9,214
33	**TOTAL ANNUAL OPERATING COSTS**						**$ 826,214**

D. PROCESSING COSTS

34	Cost of Money	@	10%	$ 235,691
35	Depreciation of Capital			$ 185,194
36	Operating Costs			$ 826,214
37	**TOTAL ANNUAL PROCESSING COSTS**			**$ 1,247,100**

38	**UNIT PROCESSING COST**	**$ 20.79**	**per tonne**	**$ 0.0208**	**per kg**
				$ 0.0094	**per lb**

Worksheet No. 1b

ESTIMATE OF GAMMA PROCESSING COST FOR A TOTE BOX IRRADIATOR ($US)

Meat & Poultry Pasteurization Scenario

A. INPUT VARIABLES

1	Cobalt	$	1.34	per Curie
2	Replenishment Cobalt	$	1.34	per Curie
3	Irradiator Efficiency		41%	
4	Annual Operating Hours		8000	hours
5	Minimum Dose		1.75	kGy
6	Throughput		60,000	tonnes/annum
7	Container Rental Charge	$	2,000.00	
8	Number of Product Handlers		16	
9	Cobalt Required		598,628	Curies
10	Replenishment Cobalt Required		73,631	Curies
11	Product Density		0.4	g/cc

B. CAPITAL ITEMS

12	Irradiator					$	900,000
13	Tote Boxes	100 totes	@		550	$	55,000
14	Auxiliary Equipment					$	160,000
15	Warehouse Conveyors	400 feet	@		700	$	280,000
16	Other						
17	Cobalt 60 Requirement (Ci)				598628 Curies		
18	Cobalt 60 Cost	598628 Ci	@	$	1.34	$	802,162
19	Container Rental	3 cont.	@	$	2,000	$	6,000
20	Land (1.5-2.5 acres)						
21	Radiation Shield					$	800,000
22	Warehouse						
23	Shipping					$	45,000
24	Start Up Cost					$	40,000
25	**TOTAL CAPITAL**					**$**	**3,088,162**

C. OPERATING ITEMS

26	Salary:	Manager					$	85,000
		RSO					$	65,000
		Operators	4	@	$	40,000.00	$	160,000
		Product handlers	16	@	$	22,000.00	$	352,000
27	Utilities and Communications						$	15,000
28	Maintenance and Repairs						$	60,000
29	Taxes and Insurance						$	20,000
30	Operational Supplies						$	30,000
31	Administration						$	30,000
32	Cobalt 60 Replenishment		73631 Ci	@	$	1.34	$	98,666
33	**TOTAL ANNUAL OPERATING COSTS**						**$**	**915,666**

D. PROCESSING COSTS

34	Cost of Money		@	10%	$	308,816
35	Depreciation of Capital				$	234,077
36	Operating Costs				$	915,666
37	**TOTAL ANNUAL PROCESSING COSTS**				**$**	**1,458,559**
38	**UNIT PROCESSING COST**	**24.31**	**per tonne**	**0.0243 per kg**		
				0.0110 per lb		

TABLE I. UNIT PROCESSING COSTS FOR A TOTE BOX IRRADIATOR
(all values are in US $ per kg)

	Dose (kGy)	Throughput (tons/annum)			
		10000	20000	60000	100000
	0.15	0.1234	0.0621		
	0.25	0.1240	0.0626		
Density 0.1 g/cc	0.50	0.1253	0.0640		
Efficiency 17%	1.00	0.1280	0.0666		
	1.75	0.1320	0.0706		
	2.50	0.1360	0.0747		
	5.00	0.1493	0.0880		
	10.00	0.1760	0.1147		
	0.15	0.1231	0.0618	0.0209	
	0.25	0.1234	0.0621	0.0212	
Density 0.2 g/cc	0.50	0.1241	0.0628	0.0219	
Efficiency 30%	1.00	0.1256	0.0643	0.0235	
	1.75	0.1279	0.0666	0.0257	
	2.50	0.1302	0.0689	0.0280	
	5.00	0.1377	0.0764	0.0356	
	10.00	0.1529	0.0916	0.0507	
	0.15	0.1230	0.0617	0.0208	0.0126
	0.25	0.1232	0.0619	0.0210	0.0129
	0.50	0.1238	0.0625	0.0216	0.0135
Density 0.3 g/cc	1.00	0.1250	0.0637	0.0228	0.0146
Efficiency 38%	1.75	0.1268	0.0655	0.0246	0.0164
	2.50	0.1286	0.0673	0.0264	0.0182
	5.00	0.1346	0.0732	0.0324	0.0242
	10.00	0.1465	0.0852	0.0443	0.0361
	0.15	0.1230	0.0616	0.0208	0.0126
	0.25	0.1232	0.0619	0.0210	0.0128
Density 0.4 g/cc	0.50	0.1237	0.0624	0.0215	0.0134
Efficiency 41%	1.00	0.1248	0.0635	0.0226	0.0145
	1.75	0.1265	0.0652	0.0243	0.0161
	2.50	0.1282	0.0668	0.0260	0.0178
	5.00	0.1337	0.0724	0.0315	0.0233
	10.00	0.1447	0.0834	0.0426	0.0344

N.B. Numbers in shaded areas derived from Worksheets 1a & 1b.

Worksheet No. 2a

ESTIMATE OF GAMMA PROCESSING COST FOR A CARRIER IRRADIATOR ($US)

Fruit & Vegetable Disinfestation Scenario

A. INPUT VARIABLES								
1	Cobalt		$	1.46 per Curie				
2	Replenishment Cobalt		$	1.46 per Curie				
3	Irradiator Efficiency			38%				
4	Annual Operating Hours			8000 hours				
5	Minimum Dose			0.15 kGy				
6	Throughput			60,000 tonnes/annum				
7	Container Rental Charge		$	2,000.00				
8	Number of Product Handlers			12				
9	Cobalt Required			55362 Curies				
10	Replenishment Cobalt Required			6810 Curies				
11	Product Density			0.4 g/cc				
B. CAPITAL ITEMS								
12	Irradiator						$	1,400,000
13	Tote Boxes	0 totes					$	-
14	Auxiliary Equipment						$	160,000
15	Warehouse Conveyors	0 feet					$	-
16	Other						$	160,000
17	Cobalt 60 Requirement (Ci)					55362 Curies		
18	Cobalt 60 Cost	55362 Ci	@		$	1.46	$	80,828
19	Container Rental	1 cont.	@		$	2,000	$	2,000
20	Land (1.5-2.5 acres)							
21	Radiation Shield						$	1,100,000
22	Warehouse							
23	Shipping						$	60,000
24	Start Up Cost						$	50,000
25	**TOTAL CAPITAL**						**$**	**3,012,828**
C. OPERATING ITEMS								
26	Salary: Manager						$	85,000
	RSO						$	65,000
	Operators	4		@	$	40,000.00	$	160,000
	Product handlers	12		@	$	22,000.00	$	264,000
27	Utilities and Communications						$	80,000
28	Maintenance and Repairs						$	55,000
29	Taxes and Insurance						$	30,000
30	Operational Supplies						$	75,000
31	Administration						$	90,000
32	Cobalt 60 Replenishment	6810 Ci		@	$	1.46	$	9,942
33	**TOTAL ANNUAL OPERATING COSTS**						**$**	**913,942**
D. PROCESSING COSTS								
34	Cost of Money			@		10%	$	301,283
35	Depreciation of Capital						$	232,589
36	Operating Costs						$	913,942
37	**TOTAL ANNUAL PROCESSING COSTS**						**$**	**1,447,813**
38	**UNIT PROCESSING COST**	**$**	**24.13 per tonne**	**$**	**0.0241 per kg**			
				$	**0.0110 per lb**			

Worksheet No. 2b

ESTIMATE OF GAMMA PROCESSING COST FOR A CARRIER IRRADIATOR ($US)

Meat Pasteurization Scenario

A. INPUT VARIABLES

1	Cobalt	$	1.34	per Curie
2	Replenishment Cobalt	$	1.34	per Curie
3	Irradiator Efficiency		38%	
4	Annual Operating Hours		8000	hours
5	Minimum Dose		1.75	kGy
6	Throughput		60,000	tonnes/annum
7	Container Rental Charge	$	2,000.00	
8	Number of Product Handlers		12	
9	Cobalt Required		645888	Curies
10	Replenishment Cobalt Required		79444	Curies
11	Product Density		0.4	g/cc

B. CAPITAL ITEMS

12	Irradiator					$	1,400,000
13	Tote Boxes	0 totes				$	-
14	Auxiliary Equipment					$	160,000
15	Warehouse Conveyors	0 feet				$	-
16	Other					$	160,000
17	Cobalt 60 Requirement (Ci)				645888 Curies		
18	Cobalt 60 Cost	645888 Ci	@	$	1.34	$	865,490
19	Container Rental	4 cont.	@	$	2,000	$	8,000
20	Land (1.5-2.5 acres)						
21	Radiation Shield					$	1,100,000
22	Warehouse						
23	Shipping					$	60,000
24	Start Up Cost					$	50,000
25	**TOTAL CAPITAL**					**$**	**3,803,490**

C. OPERATING ITEMS

26	Salary:	Manager					$	85,000
		RSO					$	65,000
		Operators	4	@	$	40,000.00	$	160,000
		Product handlers	12	@	$	22,000.00	$	264,000
27	Utilities and Communications						$	80,000
28	Maintenance and Repairs						$	55,000
29	Taxes and Insurance						$	30,000
30	Operational Supplies						$	75,000
31	Administration						$	90,000
32	Cobalt 60 Replenishment		79444 Ci	@	$	1.34	$	106,455
33	**TOTAL ANNUAL OPERATING COSTS**						**$**	**1,010,455**

D. PROCESSING COSTS

34	Cost of Money	@	10%	$	380,349
35	Depreciation of Capital			$	285,499
36	Operating Costs			$	1,010,455
37	**TOTAL ANNUAL PROCESSING COSTS**			**$**	**1,676,304**

38	**UNIT PROCESSING COST**	**$**	**27.94**	**per tonne**	**$**	**0.0279**	**per kg**
					$	**0.0127**	**per lb**

TABLE II. UNIT PROCESSING COSTS FOR A CARRIER IRRADIATOR
(all values are in US $ per kg)

	Dose (kGy)	Throughput (tons/annum)			
		10000	20000	60000	100000
	0.15	0.1433	0.0720		
	0.25	0.1438	0.0725		
Density 0.1 g/cc	0.50	0.1451	0.0738		
Efficiency 18%	1.00	0.1476	0.0763		
	1.75	0.1514	0.0801		
	2.50	0.1552	0.0839		
	5.00	0.1678	0.0965		
	10.00	0.1930	0.1217		
	0.15	0.1430	0.0717	0.0242	
	0.25	0.1433	0.0721	0.0245	
Density 0.2 g/cc	0.50	0.1441	0.0728	0.0253	
Efficiency 29%	1.00	0.1457	0.0744	0.0269	
	1.75	0.1480	0.0768	0.0292	
	2.50	0.1504	0.0791	0.0316	
	5.00	0.1582	0.0869	0.0394	
	10.00	0.1738	0.1026	0.0550	
	0.15	0.1429	0.0717	0.0241	0.0146
	0.25	0.1432	0.0719	0.0244	0.0149
	0.50	0.1439	0.0726	0.0251	0.0156
Density 0.3 g/cc	1.00	0.1452	0.0739	0.0264	0.0168
Efficiency 35%	1.75	0.1471	0.0758	0.0283	0.0188
	2.50	0.1490	0.0778	0.0302	0.0207
	5.00	0.1555	0.0842	0.0367	0.0272
	10.00	0.1685	0.0972	0.0497	0.0402
	0.15	0.1429	0.0716	0.0241	0.0146
	0.25	0.1432	0.0719	0.0244	0.0149
Density 0.4 g/cc	0.50	0.1438	0.0725	0.0250	0.0154
Efficiency 38%	1.00	0.1449	0.0737	0.0261	0.0166
	1.75	0.1467	0.0755	0.0279	0.0184
	2.50	0.1485	0.0772	0.0297	0.0202
	5.00	0.1545	0.0832	0.0357	0.0262
	10.00	0.1664	0.0952	0.0476	0.0381
	0.15	0.1429	0.0716	0.0241	0.0146
	0.25	0.1431	0.0718	0.0243	0.0148
Density 0.5 g/cc	0.50	0.1437	0.0724	0.0249	0.0154
Efficiency 40%	1.00	0.1448	0.0735	0.0260	0.0165
	1.75	0.1465	0.0752	0.0277	0.0182
	2.50	0.1482	0.0770	0.0294	0.0199
	5.00	0.1539	0.0826	0.0351	0.0256
	10.00	0.1652	0.0940	0.0464	0.0369

N.B. Numbers in shaded areas derived from Worksheets 2a & 2b

Worksheet No. 3

ESTIMATE OF GAMMA PROCESSING COST FOR A PALLET CARRIER IRRADIATOR ($US)

Fruit & Vegetable Disinfestation Scenario

A. INPUT VARIABLES

1	Cobalt	$	1.46	per Curie
2	Replenishment Cobalt	$	1.46	per Curie
3	Irradiator Efficiency		25%	
4	Annual Operating Hours		8000	hours
5	Minimum Dose		0.15	kGy
6	Throughput		60,000	tonnes/annum
7	Container Rental Charge	$	2,000.00	
8	Number of Product Handlers		6	
9	Cobalt Required		84150	Curies
10	Replenishment Cobalt Required		10350	Curies
11	Product Density		0.4	g/cc

B. CAPITAL ITEMS

12	Irradiator						$	2,000,000
13	Tote Boxes	0 totes	@		850		$	-
14	Auxiliary Equipment						$	250,000
15	Warehouse Conveyors	0 feet	@		400		$	-
16	Other						$	170,000
17	Cobalt 60 Requirement (Ci)				84150	Curies		
18	Cobalt 60 Cost	84150 Ci	@	$	1.46		$	122,859
19	Container Rental	1 cont.	@	$	2,000		$	2,000
20	Land (1.5-2.5 acres)							
21	Radiation Shield						$	1,200,000
22	Warehouse							
23	Shipping						$	60,000
24	Start Up Cost						$	55,000
25	**TOTAL CAPITAL**						**$**	**3,859,859**

C. OPERATING ITEMS

26	Salary:	Manager					$	85,000
		RSO					$	65,000
		Operators	4	@	$	40,000.00	$	160,000
		Product handlers	6	@	$	22,000.00	$	132,000
27	Utilities and Communications						$	80,000
28	Maintenance and Repairs						$	55,000
29	Taxes and Innsurance						$	30,000
30	Operational Supplies						$	85,000
31	Administartion						$	90,000
32	Cobalt 60 Replenishmnt		10350 Ci	@	$	1.46	$	15,112
33	**TOTAL ANNUAL OPERATING COSTS**						**$**	**797,112**

D. PROCESSING COSTS

34	Cost of Money	@	10%	$	385,986
35	Depreciation of Capital			$	309,891
36	Operating Costs			$	797,112
37	**TOTAL ANNUAL PROCESSING COSTS**			**$**	**1,492,988**

38	**UNIT PROCESSING COST**	**$**	**24.88**	**per tonne**	**$**	**0.0249**	**per kg**
					$	**0.0113**	**per lb**

TABLE III. UNIT PROCESSING COSTS FOR A PALLET CARRIER IRRADIATOR
(all values are in US $ per kg)

	Dose (kGy)	Throughput (tons/annum)			
		10000	20000	60000	100000
Density 0.2 g/cc Efficiency 17%	0.15	0.1466	0.0737	0.0252	0.0154
	0.25	0.1472	0.0743	0.0257	0.0160
	0.50	0.1486	0.0758	0.0272	0.0175
	1.00	0.1516	0.0787	0.0301	0.0204
	1.75	0.1559	0.0830	0.0345	0.0248
	2.50	0.1603	0.0874	0.0388	0.0291
	5.00	0.1748	0.1019	0.0534	0.0436
	10.00	0.2039	0.1310	0.0824	0.0727
Density 0.3 g/cc Efficiency23%	0.15	0.1464	0.0735	0.0249	0.0152
	0.25	0.1468	0.0739	0.0254	0.0156
	0.50	0.1479	0.0750	0.0264	0.0167
	1.00	0.1500	0.0772	0.0286	0.0189
	1.75	0.1533	0.0804	0.0318	0.0221
	2.50	0.1565	0.0836	0.0350	0.0253
	5.00	0.1672	0.0944	0.0458	0.0361
	10.00	0.1887	0.1159	0.0673	0.0576
Density 0.4 g/cc Efficiency25%	0.15	0.1463	0.0735	0.0249	0.0152
	0.25	0.1467	0.0739	0.0253	0.0156
	0.50	0.1477	0.0748	0.0263	0.0166
	1.00	0.1497	0.0768	0.0282	0.0185
	1.75	0.1527	0.0798	0.0312	0.0215
	2.50	0.1556	0.0828	0.0342	0.0245
	5.00	0.1655	0.0926	0.0441	0.0343
	10.00	0.1853	0.1124	0.0638	0.0541

N.B. Number in shaded area derived from Worksheet 3.

Worksheet No. 4

ESTIMATE OF GAMMA PROCESSING COST FOR A PALLET CONVEYOR IRRADIATOR ($US)
Fruit & Vegetable Disinfestation Scenario

A. INPUT VARIABLES								
1	Cobalt		$	1.46 per Curie				
2	Replenishment Cobalt		$	1.46 per Curie				
3	Irradiator Efficiency			29%				
4	Annual Operating Hours			8000 hours				
5	Minimum Dose			0.15 kGy				
6	Throughput			60,000 tonnes				
7	Container Rental Charge		$	2,000.00				
8	Number of Product Handlers			6				
9	Cobalt Required			72543				
10	Replenishment Cobalt Required			8923				
11	Product Density			0.4 g/cc				
B. CAPITAL ITEMS								
12	Irradiator						$	1,800,000
13	Tote Boxes		0 totes	@		850	$	-
14	Auxiliary Equipment						$	250,000
15	Warehouse Conveyors		0 feet	@		400	$	-
16	Other						$	170,000
17	Cobalt 60 Requirement (Ci)					72543 Curies	$	-
18	Cobalt 60 Cost		72543 Ci	@	$	1.46	$	105,913
19	Container Rental		1 cont.	@	$	2,000	$	2,000
20	Land (1.5-2.5 acres)							
21	Radiation Shield						$	1,000,000
22	Warehouse							
23	Shipping						$	60,000
24	Start Up Cost						$	55,000
25	**TOTAL CAPITAL**						**$**	**3,442,913**
C. OPERATING ITEMS								
26	Salary:	Manager					$	85,000
		RSO					$	65,000
		Operators	4	@	$	40,000.00	$	160,000
		Product handlers	6	@	$	22,000.00	$	132,000
27	Utilities and Communications						$	80,000
28	Maintenance and Repairs						$	55,000
29	Taxes and Insurance						$	30,000
30	Operational Supplies						$	85,000
31	Administration						$	90,000
32	Cobalt 60 Replenishment		8923 Ci	@	$	1.46	$	13,027
33	**TOTAL ANNUAL OPERATING COSTS**						**$**	**795,027**
D. PROCESSING COSTS								
34	Cost of Money			@		10%	$	344,291
35	Depreciation of Capital						$	280,761
36	Operating Costs						$	795,027
37	**TOTAL ANNUAL PROCESSING COSTS**						**$**	**1,420,079**
38	**UNIT PROCESSING COST**	**$**	**23.67 per tonne**		**$**	**0.0237 per kg**		
					$	**0.0108 per lb**		

TABLE IV. UNIT PROCESSING COSTS FOR A PALLET CONVEYOR IRRADIATOR
(all values are in US $ per kg)

	Dose (kGy)	Throughput (tonnes/annum)			
		10000	20000	60000	100000
Density 0.1 g/cc Efficiency 14%	0.15	0.1400	0.0705	0.0242	0.0150
	0.25	0.1407	0.0712	0.0249	0.0157
	0.50	0.1425	0.0730	0.0267	0.0174
	1.00	0.1460	0.0765	0.0302	0.0210
	1.75	0.1513	0.0818	0.0355	0.0263
	2.50	0.1566	0.0871	0.0408	0.0315
	5.00	0.1742	0.1048	0.0585	0.0492
	10.00	0.2096	0.1401	0.0938	0.0845
Density 0.2 g/cc Efficiency 24%	0.15	0.1396	0.0701	0.0238	0.0145
	0.25	0.1400	0.0705	0.0242	0.0149
	0.50	0.1410	0.0715	0.0252	0.0160
	1.00	0.1431	0.0736	0.0273	0.0180
	1.75	0.1461	0.0767	0.0304	0.0211
	2.50	0.1492	0.0798	0.0335	0.0242
	5.00	0.1595	0.0901	0.0438	0.0345
	10.00	0.1801	0.1107	0.0643	0.0551
Density 0.4 g/cc Efficiency 29%	0.15	0.1395	0.0700	0.0237	0.0144
	0.25	0.1398	0.0703	0.0240	0.0147
	0.50	0.1406	0.0712	0.0249	0.0156
	1.00	0.1423	0.0729	0.0266	0.0173
	1.75	0.1449	0.0754	0.0291	0.0199
	2.50	0.1475	0.0780	0.0317	0.0224
	5.00	0.1560	0.0865	0.0402	0.0309
	10.00	0.1730	0.1036	0.0572	0.0480

N.B. Number in shaded area derived from Worksheet 4.

Worksheet No.5

ESTIMATE OF GAMMA PROCESSING COST FOR A HORIZONTAL SOURCE IRRADIATOR ($US)

Meat Pasteurization Scenario

A. INPUT VARIABLES						
1	Cobalt	$	1.34 per Curie			
2	Replenishment Cobalt	$	1.34 per Curie			
3	Irradiator Efficiency		33%			
4	Annual Operating Hours		8000 hours			
5	Minimum Dose		1.75 kGy			
6	Throughput		60,000 tonnes			
7	Container Rental Charge	$	6,000.00			
8	Number of Product Handlers		6			
9	Cobalt Required		743750			
10	Replenishment Cobalt Required		91481			
11	Product Density		0.4			
B. CAPITAL ITEMS						
12	Irradiator				$	2,700,000
13	Tote Boxes	0 totes @	$	850.00		
14	Auxiliary Equipment				$	450,000
15	Warehouse Conveyors	0 feet @	$	400.00		
16	Other					
17	Cobalt 60 Requirement (Ci)			743750 Curies		
18	Cobalt 60 Cost	743750 Ci @	$	1.34	$	996,625
19	Container Rental	4 cont. @	$	2,000	$	8,000
20	Land (1.5-2.5 acres)					
21	Radiation Shield				$	1,600,000
22	Warehouse					
23	Shipping				$	50,000
24	Start Up Cost				$	40,000
25	**TOTAL CAPITAL**				**$**	**5,844,625**
C. OPERATING ITEMS						
26	Salary: Manager				$	85,000
	RSO				$	65,000
	Operators	4	@ $	40,000.00	$	160,000
	Product handlers	6	@ $	22,000.00	$	132,000
27	Utilities and Communications				$	45,000
28	Maintenance and Repairs				$	40,000
29	Taxes and Insurance				$	30,000
30	Operational Supplies				$	85,000
31	Administration				$	90,000
32	Cobalt 60 Replenishment	91481 Ci	@ $	1.34	$	122,585
33	**TOTAL ANNUAL OPERATING COSTS**				**$**	**854,585**
D. PROCESSING COSTS						
34	Cost of Money		@	10%	$	584,463
35	Depreciation of Capital				$	455,242
36	Operating Costs				$	854,585
37	**TOTAL ANNUAL PROCESSING COSTS**				**$**	**1,894,289**

38 **UNIT PROCESSING COST** **$ 31.57 per tonne** **$ 0.0316 per kg**

$ 0.0144 per lb

TABLE V. UNIT PROCESSING COSTS FOR A HORIZONTAL SOURCE IRRADIATOR
(all values are in US $ per kg)

	Dose (kGy)	Throughput (tons/annum)			
		10000	20000	60000	
	0.15	0.1610	0.0808	0.0272	
	0.25	0.1614	0.0811	0.0276	
Density 0.3 g/cc	0.50	0.1622	0.0819	0.0284	
Efficiency 28%	1.00	0.1638	0.0835	0.0300	
	1.75	0.1662	0.0860	0.0324	
	2.50	0.1687	0.0884	0.0349	
	5.00	0.1768	0.0965	0.0430	
	10.00	0.1930	0.1127	0.0592	
	0.15	0.1610	0.0807	0.0272	
	0.25	0.1612	0.0810	0.0274	
Density 0.4 g/cc	0.50	0.1619	0.0817	0.0281	
Efficiency 33%	1.00	0.1633	0.0830	0.0295	
	1.75	0.1654	0.0851	0.0316	
	2.50	0.1674	0.0872	0.0336	
	5.00	0.1743	0.0940	0.0405	
	10.00	0.1881	0.1078	0.0543	
	0.15	0.1609	0.0806	0.0271	0.0164
	0.25	0.1612	0.0809	0.0274	0.0167
	0.50	0.1618	0.0815	0.0280	0.0172
Density 0.6 g/cc	1.00	0.1629	0.0827	0.0291	0.0184
Efficiency 38%	1.75	0.1647	0.0845	0.0309	0.0202
	2.50	0.1665	0.0862	0.0327	0.0220
	5.00	0.1725	0.0922	0.0387	0.0280
	10.00	0.1844	0.1042	0.0506	0.0399
	0.15	0.1609	0.0806	0.0271	0.0164
	0.25	0.1612	0.0809	0.0274	0.0167
Density 0.8 g/cc	0.50	0.1618	0.0815	0.0280	0.0173
Efficiency 37%	1.00	0.1630	0.0827	0.0292	0.0185
	1.75	0.1649	0.0846	0.0311	0.0203
	2.50	0.1667	0.0864	0.0329	0.0222
	5.00	0.1728	0.0925	0.0390	0.0283
	10.00	0.1851	0.1048	0.0513	0.0406

N.B. Number in shaded area derived from Worksheet 5.

5. EFFECT OF THROUGHPUT ON COSTS

Figure 6 depicts the effect of economies of scale. The examples are chosen at the dose of 0.15 kGy (0.015 Mrad) and the density of 0.4 g/cm^3. As expected, unit processing costs in all four types of irradiators decrease rapidly with initial increase of throughput. They stabilize once the minimum economic throughput is reached and surpassed. These curves extend until a particular machine's mechanical throughput limits are reached. In the case of pallet and carrier irradiators, mechanical limits usually do not represent a challenge in terms of food handling requirements. A successful irradiator will constantly operate at levels exceeding minimum economic throughputs as determined by the product owner, taking advantage of economies of scale. It is therefore imperative to correctly size the irradiator for the anticipated processing scenario.

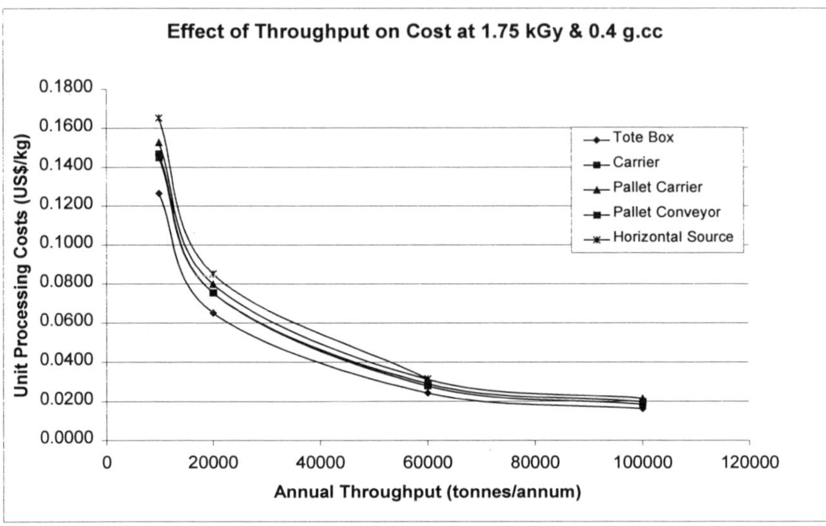

Figure 6.

6. EFFECT OF DOSE ON COSTS

The effect of increasing dose on unit processing costs is, as expected, linear, since unit processing costs directly relate to the facility's operating costs and inversely to annual throughputs. Figure 7 shows the effects of dose at the density of 0.4 g/cm^3 and a throughput of 60 million kg/annum. Within the fruit disinfestation dose range of 0.15-0.75 kGy (15-75 krad), there is no dramatic difference between the trends in various types of irradiators. However, when this diagram is extended to doses ranging up to 5 kGy (500 krad), some interesting trends emerge. In general, the conveyor concepts, due to their `overlapping product' method of irradiation, offer better cobalt utilization efficiency and, therefore, better economies. The pallet conveyor concept is most economical at lower dose ranges and breaks even at mid-range. Theoretically, at higher dose ranges, the tote box concept would be more economical. In this particular example, however, the tote box irradiator has reached its mechanical limit approximately at the point where it would be more economical, on a per unit basis, than the pallet conveyor concept.

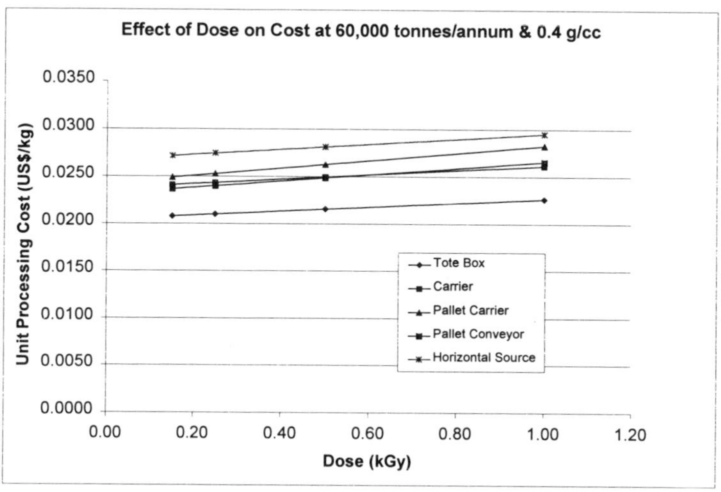

Figure 7.

A similar comparison can be made between the two carrier concepts, i.e. the pallet irradiator is more economical at lower doses and breaks even at medium doses whereas in high dose ranges the carrier irradiator is more cobalt efficient and cost effective. The pallet concept will, nevertheless, maintain its advantage of low labour requirements, higher throughput capacity and ease of loading and unloading.

It should be noted that the trends depicted in Figures 7 and 8 are tempered by the fact that our calculations assume that the cobalt source strength exactly matches the dose and throughput requirements in all cases. Realistically, source strengths are not this flexible, due to the fact that cobalt sources are supplied in discreet quantities and due to the constant decay of cobalt 60. On the other hand, for simplicity, our calculations are based on single product scenarios. Additional products that will be treated in the same facility, will result in additional cobalt 60 requirements and in more realistic considerations of cobalt 60 utilization efficiencies. Ultimately, if source strength is considered to be a constant, the effect of dose on unit processing costs will be linear and directly dependent on processing time.

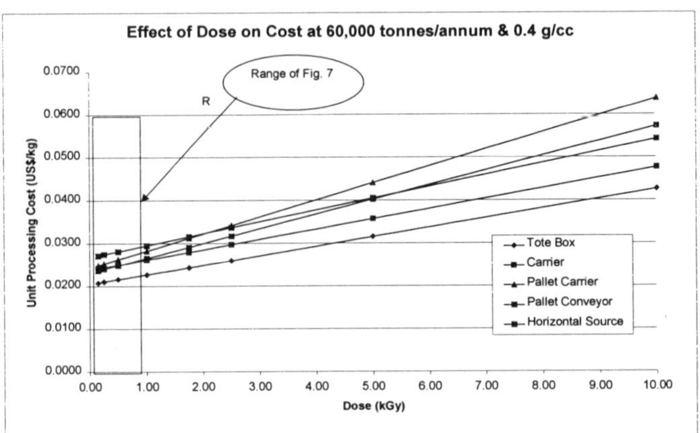

Figure 8.

7. EFFECT OF PACKING DENSITY ON COSTS

Figures 9 to 13 show the effect of various packing densities on dose-cost relationships at the constant throughput of 60,000 tonnes/annum. The effect is completely predictable, as it is controlled by the effect of product density on the irradiator's cobalt utilization efficiency. Cobalt utilization efficiency increases with increased density, up to a point where the products' self-shielding causes the utilization efficiency to decline. This turn-around point varies from unit to unit, depending on specific design features. Figure 9 shows some mechanical limitations of a tote box irradiator. At a packing density of 0.2 g/cm^3, it is capable of processing a maximum of 60,000 tonnes/annum and cannot process a product packed at or over a density of 0.5 g/cm^3 without structural modifications. Figure 10 shows that cobalt utilization efficiency of a carrier irradiator improves at high densities. This is demonstrated by lower unit processing costs. Figure 11 is similar to Figure 10 but is for a pallet irradiator. The cobalt utilization efficiency for a pallet irradiator peaks around 0.4 g/cm^3 and then starts to drop. This results in equal unit processing costs at densities of 0.3 and 0.5 g/cm^3. In the case of a pallet conveyor irradiator, equal costs are found at densities of 0.3 and 0.4 g/cm^3 (Figure 12). Similar relations exist for the horizontal source irradiator (Figure 13).

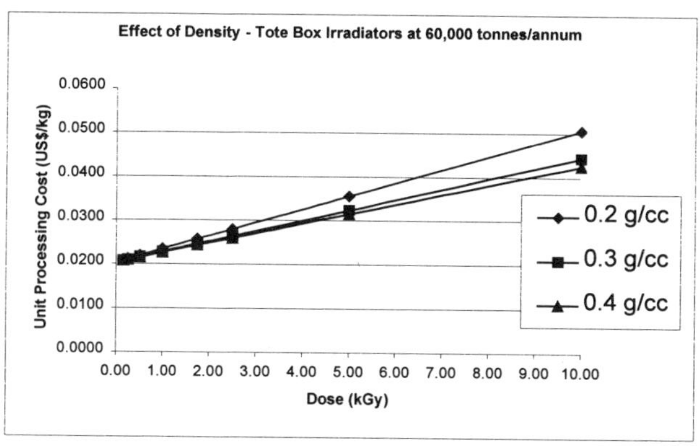

Figure 9.

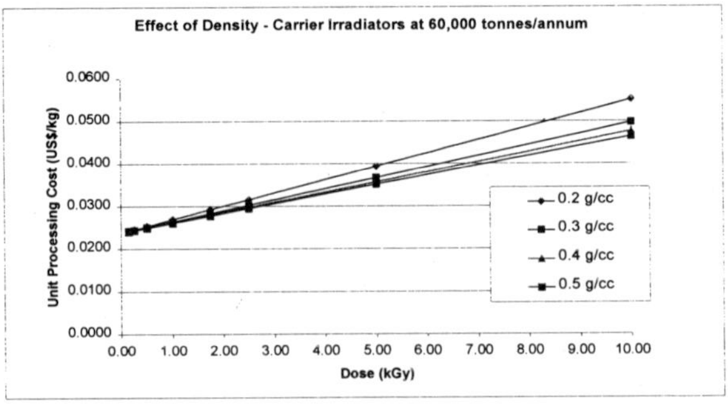

Figure 10.

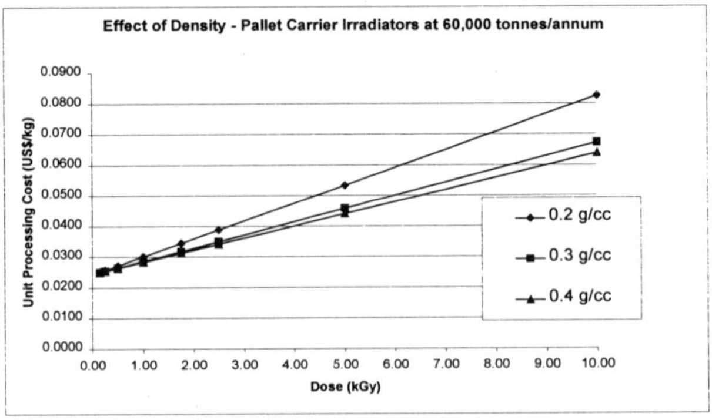

Figure 11.

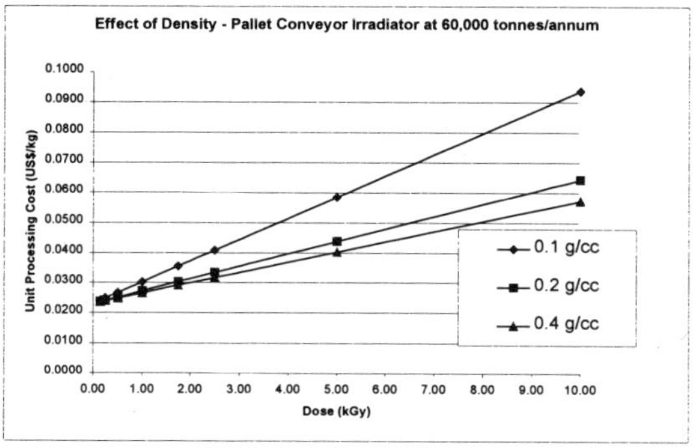

Figure 12

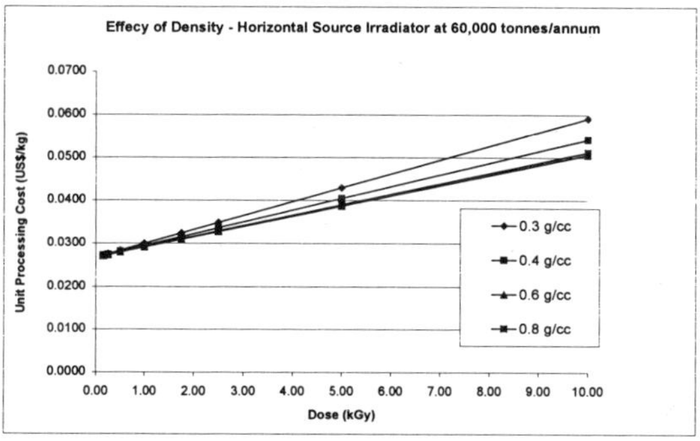

Figure 13.

8. EFFECT OF PACKING DENSITY ON COBALT UTILIZATION EFFICIENCY

Cobalt utilization efficiency increases with increased packing density up to a point of optimum efficiency and then declines. This phenomenon is caused by improved absorption efficiency with increased product density, up to a point where the optimum is reached. Beyond this point, self-absorption as a result of increased product density becomes a significant factor, preventing the radiation from reaching portions of the product stack. Density-efficiency (D-E) curves are equipment design specific. Each individual irradiator will have its own unique performance table and its own unique D-E curve. The D-E curves for the five irradiator concepts used in this paper are shown in Figure 14. Figure 15 shows the curve for another carrier irradiator of interest because it covers the full range of densities usually found in foods.

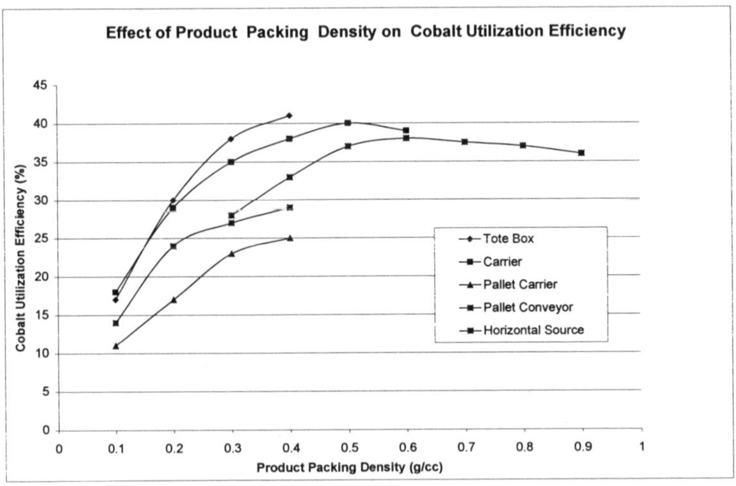

Figure 14.

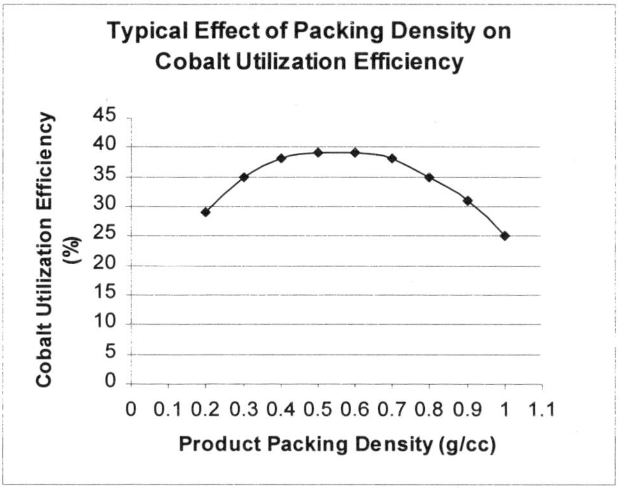

Figure 15.

9. SUMMARY

Costs of irradiation are influenced by a large number of factors. There is an immense variety of food products, applied doses, desired effects, packing densities, throughputs, special handling, packaging and logistical requirements, and local conditions to be considered in selecting food irradiation equipment and establishing unit processing costs. In this paper, we have presented examples of approaches that can be used in considering the economics of food irradiation, rather than trying to arrive at fixed figures. We hope that we have succeeded in presenting an understandable method whereby a potential user in conjunction with an equipment supplier can arrive at reasonable conclusions about specific radiation processing costs for a specific situation. Many of the inputs required for the unit processing cost worksheets can only be obtained from equipment and Cobalt 60 suppliers after the specific optimum irradiator design has been determined. In order to optimize processing costs, one must fully understand the processing scenario and design an irradiation facility that best matches it.

Economics of Machine Sources for Irradiation of Food

M. R. CLELAND, A. S. HERER and A. COKRAGAN
Ion Beam Applications
Louvain-la-Neuve, Belgium

ABSTRACT

Machine sources of ionizing energy for the irradiation of food include beams of energetic electrons and X rays (bremsstrahlung) generated when the electrons are intercepted with metallic targets. Electron energy requirements depend on the thickness and density of the products and cover a wide range from 0. 1 to 10 MeV. The lowest electron energies can be used for treating surfaces, fluids and granular materials while the highest energies can be used for fruits, vegetables and meats with thickness up to about 10 cm. X rays are capable of treating containers of such products which are more than 50 cm in thickness. Electron beam power requirements depend on the absorbed dose and the throughput rate of the treatment process and also cover a wide range from less than 10 to more than 200 kW. The dose for various applications can be as low as 0. 15 and as high as 30 kGy while the throughput rate can range from a few to hundreds of tonnes per hour. With such a variety of process specifications, the cost of irradiating food also varies substantially from one application to another. In this paper, estimates are given for the fixed and variable costs for typical applications. Fixed costs include capital amortization, utilities, maintenance and services for the facility (other than the irradiation equipment) and administration. Variable costs include utilities and maintenance for the irradiation equipment, labor and supervision for the treatment process, dosimetry, and consumable supplies. The total cost per unit of product decreases as the throughput rate increases because the fixed costs are then spread over a larger number of units.

INTRODUCTION

Machine sources of ionizing energy for the irradiation of food include beams of energetic electrons and X rays (bremsstrahlung) generated when the electrons are intercepted with metallic targets. Electron energy requirements increase with the thickness and density of the products and cover a wide range from 0.1 million to 10 million electron volts (MeV). The lower electron energies can be used for treating surfaces, fluids and granular materials while the higher energies can be used for fruits, vegetables and meats with thickness up to about 10 cm. X rays are capable of treating containers of food products which are more than 50 cm in thickness.

Electron beam power requirements increase with the absorbed dose and the throughput rate of the treatment process and also cover a wide range from a few kilowatts (kW) to more than 200 kW. The dose for various food applications can be as low as 0.15 kilograys (kGy) and as high as 30 kGy while the throughput rate can range from a few tonnes to hundreds of tonnes per hour.

With such a variety of process specifications, the cost of irradiating food also varies substantially from one application to another. In this paper, estimates are given for the fixed and variable costs for typical applications, such as the disinfestation of grain, fruits or vegetables and the disinfection of fresh meat, poultry or seafood. Fixed costs include capital amortization, supervision and maintenance labor, supplies and other expenses for the facility. Variable costs include plant labor, utilities and services, supplies and maintenance for the irradiation equipment. The unit treatment costs decrease as the throughput increases because the fixed costs are then divided by a larger number of product units.

RHODOTRON ACCELERATORS

Rhodotron electron accelerators can produce electron energies up to 10 MeV with beam powers up to 200 kW. This combination of high energy and high power is unique among the variety of accelerators that are used for radiation processing of materials and commercial products. High electron energy permits the treatment of thick products and high beam power provides high throughput rates. Rhodotrons are also suitable for generating X rays to treat products that are too thick to be done with electrons because of their limited penetration. An attractive feature of using a machine source of radiation is that either electron-beam or X-ray processing can be done. This broadens the variety of products that can be accommodated with the same equipment.

A Rhodotron accelerates electrons in a large, coaxial resonant cavity which is energized with radio-frequency (rf) power. The electrons make multiple passes through the evacuated cavity to gain high energy. The pattern of electron paths is shown by the diagram in Figure 1. The electrons are injected in short bunches at the proper phase to be accelerated by the alternating electric field. The rf field reverses polarity as the electrons pass through the central tube and it reverses again as they pass through the deflection magnets. Each pass across a radius of the cavity gives an energy gain of 0.5 MeV and a pass across the full diameter gives an energy gain of 1 MeV. With 9 deflection magnets and 10 passes, the final energy is 10 MeV. Lower energies are obtained by reducing the number of passes (1, 2).

Ion Beam Applications (IBA) now makes three models, designated TT100, TT200 and TT300. These have average beam power ratings of 35 kW, 80 kW and 150 kW, respectively. A photograph of the TT300 is shown in Figure 2. A 200 kW beam power capability has been demonstrated recently with this model (3). The resonant cavity has a diameter of 2.0 m and a height of 1.8 m. The overall diameter, including the beam deflection magnets, is 3.0 m. The resonant frequency is 107.5 megahertz (MHz). This relatively low frequency permits the use of a tetrode tube instead of a klystron for the output stage of the rf amplifier. Tetrodes are more efficient and less expensive than klystrons, which must be used with industrial linear accelerators (linacs). The low frequency and low energy gain per pass enable a Rhodotron to be operated in the continuous-wave (cw) mode with less than 100 kW of rf power dissipated in the resonant cavity. Because of this characteristic, the electron beam is essentially continuous and can be scanned at the fast sweep rates that are needed for high-power, high-throughput applications.

ELECTRON BEAM PROCESSING

The penetration of an electron beam in treated products is directly proportional to the incident electron energy and inversely proportional to the product density. It is also affected by the atomic composition of the material. The absorbed dose inside the product increases to a depth of about half the maximum electron range and then decreases from there to the end of the range. The optimum thickness for treatment from one side is the depth where the decreasing dose equals the entrance dose. The optimum thickness for treatment from opposite sides is about 2.5 times greater because the decreasing dose distributions overlap in the middle of the product. With 10 MeV electrons in fresh foods with high water content, the optimum thickness for one-sided treatment is about 3.6 cm and for treatment from opposite sides it is about 9 cm. The internal dose uniformity improves as the electron energy increases. The max/min dose ratios in fresh foods are about 1.6 at 3 MeV, 1.5 at 5 MeV and 1.3 at 10 MeV, assuming optimum thicknesses for each energy (4).

The processing rate is directly proportional to the electron beam power and inversely proportional to the absorbed dose. The absorbed dose is defined as the energy absorbed per unit mass. The international (SI) unit of dose is the gray (Gy), which is 1 J/kg. A more common unit for radiation processing is the kGy, which is 1 kJ/kg or 1 kWs/kg. According to these definitions, the absorption of 1 kW of beam power in products moving through the beam with a throughput rate of 1 kg/s produces an average dose of 1 kGy.

The beam power utilization efficiency is the ratio of the absorbed power to the emitted power. The effective beam power utilization is usually calculated with the surface or minimum dose rather than the average dose. With the optimum product thickness, the surface dose will be lower than the average dose. The higher dose inside the product and the part of the electron beam that does not strike the product account for the lost beam power. With products that have nearly uniform thickness and are close together on the conveyor, the effective beam power utilization can be in the range of 40% to 60%. For example, with 150 kW of beam power, 50% power utilization and a minimum dose of 1 kGy, the throughput rate would be 75 kg/s.

X-RAY PROCESSING

X rays are a form of electromagnetic radiation which is emitted when energetic electrons pass near atomic nuclei. The technical term is bremsstrahlung (braking radiation) and the common name is X rays. The continuous energy distribution of X-ray photons extends from a few keV up to the maximum energy of the incident electrons. The generating efficiency increases with the atomic number of the target material. With electron energies above 2 MeV, X ray emission is concentrated in the direction of the electron beam and the angular divergence decreases as the electron energy increases (5). Higher energies provide increased generating efficiency, greater penetration and improved dose uniformity in products treated with X rays. Regulations for food processing limit the maximum energy to 5 MeV so that radioactivity from photonuclear reactions will be insignificant. Recent reviews have shown that a maximum energy of 7.5 MeV also would be safe (6, 7). IBA is preparing a petition to the US Food and Drug Administration requesting that their regulations be changed to increase the maximum X-ray energy from 5 MeV to 7.5 MeV.

Calculations with Monte Carlo codes have shown that the efficiencies for X-ray generation using a tantalum target with optimum thickness are about 8% at 5 MeV and 13% at 7.5 MeV. Products are treated from opposite sides to obtain good dose uniformity. The optimum product thicknesses for maximum efficiency of X-ray utilization in water or fresh food are 34 cm at 5 MeV and 38 cm at 7.5 MeV. The max/min dose ratio with these thicknesses is about 1.5 with both energies (8, 9). Lower max/min dose ratios can be obtained with thinner packages and more than one row of product carriers passing through the treatment zone.

As with electron beam treatment, the processing rate with X rays is directly proportional to the emitted power and inversely proportional to the absorbed dose. Both the generating efficiency and the utilization efficiency must be known to calculate the mass throughput rate. With a suitable product handling system, the utilization efficiency with X-ray processing can be about the same as with electron beam processing. For example, with 150 kW of electron beam power on a tantalum target, 8% X-ray conversion efficiency, 50% utilization efficiency and a minimum dose of 1 kGy, the throughput rate would be 150 x 0.08 x 0.50 / 1.0 = 6.0 kg/s. If the minimum dose were 1.5 kGy, sufficient for pasteurizing fresh meat, then the throughput rate would be 4.0 kg/s or 240 kg/min or 14,400 kg/h.

APPLICATIONS

The use of ionizing energy to control insects in stored grains and fresh fruits and vegetables has been studied extensively. The many varieties of such pests have different radiation sensitivities, and the many varieties of agricultural commodities have different tolerances to radiation treatment. In general, the doses required to kill adult insects would have deleterious effects on the quality of fresh fruits. Therefore, the Animal and Plant Health Inspection Service (APHIS) of the US Department of Agriculture has established the policy of preventing the emergence of adult insects or sterilizing the adults already present. With fruit flies, this can be accomplished with a minimum dose of 0.25 kGy, which is tolerated by most fruits (10, 11). This dose also will prevent the proliferation of most insects in stored grain (12).

The use of ionizing energy to inactivate microorganisms in fresh meat and poultry is also well known. The Food Safety and Inspection Service (FSIS) of the US Department of Agriculture has established a minimum dose of 1.5 kGy for the treatment of fresh or frozen uncooked poultry products (13). Their proposed rule for the treatment of refrigerated or frozen uncooked meat from domestic animals does not state a minimum dose (14). However, a treatment with 1.5 kGy is sufficient to protect consumers from most food-borne illnesses.

ECONOMICS

The cost of treating food with high-energy electrons or X rays depends on the equipment specifications and the process requirements. Approximate capital costs are given in Figure 3 and operating costs are given in Tables 1, 2 and 3 for the three Rhodotron models which are described above. These figures are estimates, not firm prices, which might be different because of local conditions. The three applications chosen for these analyses are: disinfestation of grain with electrons, disinfestation of fruit and disinfection of meat with X rays. The annual production quantities and unit processing costs have been calculated in each table for operating schedules of 1, 2 and 3 shifts per day or 2,000, 4,000 and 6,000 hours per year.

The electron energy is in the range of 5 MeV to 10 MeV and the electron beam power utilization is assumed to be 50%. The maximum X-ray energy is 5 MeV, the electron to X-ray power conversion efficiency is assumed to be 8% and the X-ray power utilization is assumed to be 50%. The total capital costs have been amortized over 20 years at 8% interest. No credit has been taken into account for interest and depreciation deductions from taxable income. Labor costs are typical of western European countries and the USA. Electricity costs are based on $0.10 per kilowatt hour, which is higher than the national average in the USA.

Comparisons of the results given in each table indicate that the unit processing costs decrease as the operating time and annual production increase. This trend is caused by the relatively high fixed operating costs. Comparisons of the results given in different tables indicate that the unit processing costs decrease as the beam power of the electron accelerator increases. This trend occurs because the capital costs do not increase in direct proportion to the beam power. With full utilization of the beam power from these machines, the unit processing costs are only a few cents per kilogram for disinfecting meat with X rays. The unit costs for disinfesting fruit with X rays are lower because of the lower dose and the unit costs for disinfesting grain with electrons are much lower because of the higher processing rates with electrons in comparison to X rays.

CONCLUSION

Rhodotron accelerators can now provide high-power electron beams with high electron energies. These characteristics are suitable for processing a variety of foods with electrons or with X rays (bremsstrahlung). The high capital costs of accelerator facilities are justified by their high productivity. When such facilities are used to their full capacities, the unit processing costs are much lower than the commercial values of agricultural products.

REFERENCES

(1) Pottier, J. 1989. "A New Type of RF Electron Accelerator: The Rhodotron," Nuclear Instruments and Methods in Physics Research, B40/41: 943-945.

(2) Jongen, Y., Abs, M., Genin, F., Nguyen, A.,Capdevila, J.M. and Defrise, D. 1993. "The Rhodotron, a New 10 MeV, 100 kW, CW Metric Wave Electron Accelerator," Nuclear Instruments and Methods in Physics Research, B79:865-870.

(3) Watanabe, T. 1999. "Best Use of High-Voltage, High-Powered Electron Beams: A New Approach to Contract Irradiation Services," Proceedings of the 11th International Meeting on Radiation Processing, to be published in Radiation Physics and Chemistry.

(4) ASTM E 1649-94. "Standard Practice for Dosimetry in an Electron Beam Facility for Radiation Processing at Energies between 300 keV and 25 MeV," 12.02, Annual Book of ASTM Standards, American Society for Testing and Materials, West Conshohocken, PA 19428, USA.

(5) ASTM E 1608-94. "Standard Practice for Dosimetry in an X-Ray (Brems-strahlung) Facility for Radiation Processing," 12.02, Annual Book of ASTM Standards, American Society for Testing and Materials, West Conshohocken, PA 19428, USA.

(6) Anon. 1995. "Report of the Consultant's Meeting on the Development of X-Ray Machines for Food Irradiation," Food and Agriculture Organization, International Atomic Energy Agency, A-1400 Vienna, Austria.

(7) Brynjolfsson, A. 1999. "Natural and Induced Radioactivity in Food," Proceedings of the Global '99 Conference on Future Nuclear Systems, American Nuclear Society, La Grange Park, IL 60526 USA.

(8) Meissner, J., Abs, M., Cleland, M.R., Herer, A.S., Jongen, Y., Kuntz, F. and Strasser, A. 1999. "X-Ray Treatment at 5 MeV and Above," Proceedings of the 11th International Meeting on Radiation Processing, Melbourne, Australia, March 14-19, 1999, to be published in Radiation Physics and Chemistry.
(9) Cleland, M.R., Herer, A.S., Meissner, J., Abs, M., Jongen, Y., Kuntz, F. and Strasser, A. 1999. "X-Ray Technology for Food Processing," Proceedings of the Intertech Conference, Food Irradiation 99, May 12-14, 1999, Arlington, VA USA
(10) Anon. 1996. "The Application of Irradiation to Phytosanitary Problems," Notice of Policy, Animal and Plant Health Inspection Service, US Department of Agriculture, 7 CFR Part 319, Federal Register, 61(95): 24433-24439, May 15, 1996.
(11) Hallman, G.J. 1999. "Ionizing Radiation Quarantine Treatments Against Tephritid Fruit Flies," Postharvest Biology and Technology, 16: 93-106.
(12) Tilton, E.W. and Burditt, Jr., A.K. Chapter 7, "Insect Disinfestation of Grain and Fruit," Josephson, E.S. and Peterson, M.S. 1983. Preservation of Food by Ionizing Radiation, Boca Raton, FL: CRC Press, Inc. pp. 215-229.
(13) Anon. 1992. "Irradiation of Poultry Products," Final Rule, Food Safety and Inspection Service, US Department of Agriculture, 9 CFR Part 381, Federal Register 57(183): 43588-43600, September 21, 1992.
(14) Anon. 1999. "Irradiation of Meat and Meat Products," Proposed Rule, Food Safety and Inspection Service, US Department of Agriculture, 9 CFR Parts 317, 318 and 381, Federal Register 64(36): 9089-9105, February 24, 1999.

TABLE 1. IRRADIATION PROCESS COSTS

Electron Accelerator - IBA Rhodotron TT100
Electron Energy = 5 to 10 MeV, Beam Power = 35 kW

FACILITY OPERATING COSTS IN $ USA/YR

Operating Schedule in Hours per Year	2,000	4,000	6,000
Fixed Operating Costs			
Capital Amortization, 20 yr at 8% interest	524,539	524,539	524,539
Supervision and Maintenance Labor	150,000	150,000	150,000
Supplies and Other Expenses	80,000	80,000	80,000
Total Fixed Operating Costs	**754,539**	**754,539**	**754,539**
Variable Operating Costs			
Plant Labor	130,000	260,000	390,000
Utilities and Services	60,000	120,000	180,000
Supplies and Maintenance	40,000	80,000	120,000
Total Variable Operating Costs	**230,000**	**460,000**	**690,000**
Total Fixed +Variable Operating Costs	**984,539**	**1,214,539**	**1,444,539**

UNIT PROCESSING COSTS IN $ USA/TONNE

Disinfestation of Grain with Electrons			
Production in tonnes/yr at 0.25 kGy	504,000	1,008,000	1,512,000
Unit Processing Cost in $/tonne USA	**1.95**	**1.20**	**0.96**
Disinfestation of Fruit with X Rays			
Production in tonnes/yr at 0.25 kGy	40,320	80,640	120,960
Unit Processing Cost in $/tonne USA	**24.42**	**15.06**	**11.94**
Disinfection of Meat with X Rays			
Production in tonnes/yr at 1.5 kGy	6,720	13,440	20,160
Unit Processing Cost in $/tonne USA	**146.51**	**90.37**	**71.65**

Figure 1. Diagram showing the multiple passes of an electron beam through a Rhodotron accelerator cavity.

TABLE 2. IRRADIATION PROCESS COSTS

Electron Accelerator - IBA Rhodotron TT200
Electron Energy = 5 to 10 MeV, Beam Power = 80 kW

FACILITY OPERATING COSTS IN $ USA/YR

Operating Schedule in Hours per Year	2,000	4,000	6,000
Fixed Operating Costs			
Capital Amortization, 20 yr at 8% interest	718,058	718,058	718,058
Supervision and Maintenance Labor	150,000	150,000	150,000
Supplies and Other Expenses	110,000	110,000	110,000
Total Fixed Operating Costs	**978,058**	**978,058**	**978,058**
Variable Operating Costs			
Plant Labor	260,000	520,000	780,000
Utilities and Services	90,000	180,000	270,000
Supplies and Maintenance	60,000	120,000	180,000
Total Variable Operating Costs	**410,000**	**820,000**	**1,230,000**
Total Fixed +Variable Operating Costs	**1,388,058**	**1,798,058**	**2,208,058**

UNIT PROCESSING COSTS IN $ USA/TONNE

Disinfestation of Grain with Electrons			
Production in tonnes/yr at 0.25 kGy	1,152,000	2,304,000	3,456,000
Unit Processing Cost in $/tonne USA	**1.20**	**0.78**	**0.64**
Disinfestation of Fruit with X Rays			
Production in tonnes/yr at 0.25 kGy	92,160	184,320	276,480
Unit Processing Cost in $/tonne USA	**15.06**	**9.76**	**7.99**
Disinfection of Meat with X Rays			
Production in tonnes/yr at 1.5 kGy	15,360	30,720	46,080
Unit Processing Cost in $/tonne USA	**90.37**	**58.53**	**47.92**

Figure 2. A photograph of the Rhodotron electron accelerator made by Ion Beam Applications (IBA) in Louvain-la-Neuve, Belgium.

TABLE 3. IRRADIATION PROCESS COSTS

Electron Accelerator - IBA Rhodotron TT300
Electron Energy = 5 to 10 MeV, Beam Power = 150 kW

FACILITY OPERATING COSTS IN $ USA/YR

Operating Schedule in Hours per Year	2,000	4,000	6,000
Fixed Operating Costs			
Capital Amortization, 20 yr at 8% interest	972,689	972,689	972,689
Supervision and Maintenance Labor	150,000	150,000	150,000
Supplies and Other Expenses	120,000	120,000	120,000
Total Fixed Operating Costs	**1,242,689**	**1,242,689**	**1,242,689**
Variable Operating Costs			
Plant Labor	260,000	520,000	780,000
Utilities and Services	120,000	240,000	360,000
Supplies and Maintenance	60,000	120,000	180,000
Total Variable Operating Costs	**440,000**	**880,000**	**1,320,000**
Total Fixed +Variable Operating Costs	**1,682,689**	**2,122,689**	**2,562,689**

UNIT PROCESSING COSTS IN $ USA/TONNE

Disinfestation of Grain with Electrons			
Production in tonnes/yr at 0.25 kGy	2,160,000	4,320,000	6,480,000
Unit Processing Cost in $/tonne USA	**0.78**	**0.49**	**0.40**
Disinfestation of Fruit with X Rays			
Production in tonnes/yr at 0.25 kGy	172,800	345,600	518,400
Unit Processing Cost in $/tonne USA	**9.74**	**6.14**	**4.94**
Disinfection of Meat with X Rays			
Production in tonnes/yr at 1.5 kGy	28,800	57,600	86,400
Unit Processing Cost in $/tonne USA	**58.43**	**36.85**	**29.66**

Figure 3. Chart showing the trends of capital cost components vs electron beam power ratings for the three Rhodotron models, TT100, TT200 and TT300.

A Accelerator and IBA Supplies	1 = A
B Related Accelerator Equipment	2 = A + B
C Processing Equipment	3 = A + B + C
D Professional Support and Services	4 = A + B + C + D
E Contingencies	5 = A + B + C + D + E

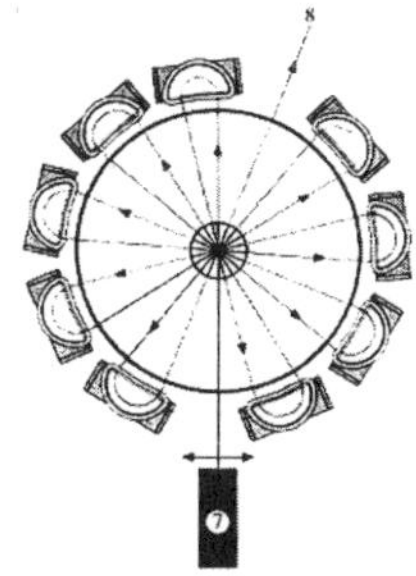

Figure 1. Diagram showing the multiple passes of an electron beam through a Rhodotron accelerator cavity.

Figure 2. A photograph of the Rhodotron electron accelerator made by Ion Beam Applications (IBA) in Louvain-la-Neuve, Belgium.

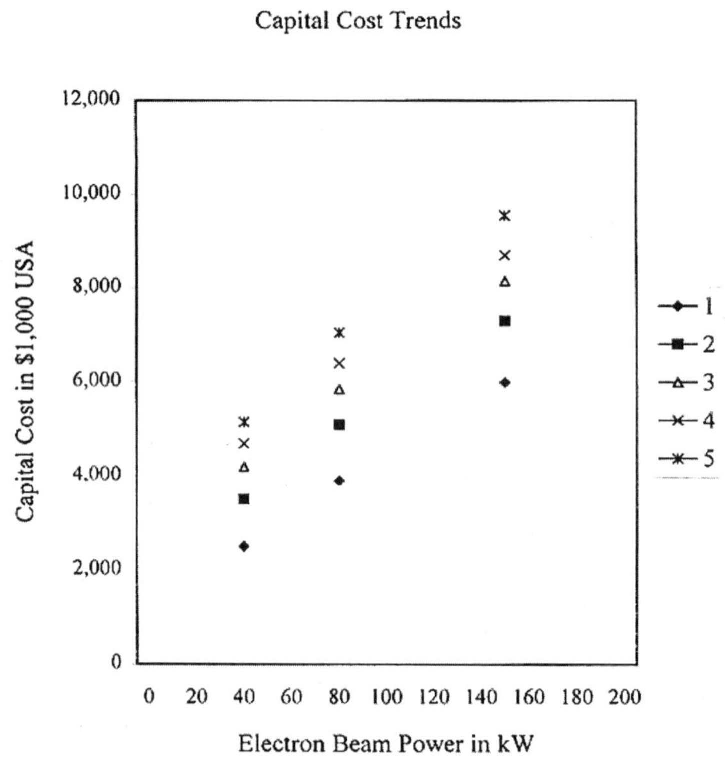

Figure 3. Chart showing the trends of capital cost components vs electron beam power ratings for the three Rhodotron models, TT100, TT200 and TT300.

A	Accelerator and IBA Supplies	1 = A
B	Related Accelerator Equipment	2 = A + B
C	Processing Equipment	3 = A + B + C
D	Professional Support and Services	4 = A + B + C + D
E	Contingencies	5 = A + B + C + D + E

United States Consumer Choice of Irradiated Food

C. M. BRUHN
Director, Center for Consumer Research
University of California, Davis, CA, USA

ABSTRACT

Consumers respond to benefits. Irradiation offers advantages not obtainable by other processing methods. More consumers have become aware of these advantages and have heard widespread endorsement by health authorities. Plans are proceeding for wide scale marketing of irradiated food in at least one state. Media coverage of illness from *E. coli* 0157:H7, *Salmonella sp.*, and other pathogens has increased consumer concern about microbiological hazards in food. More people volunteer concerns about microbiological hazards than any other potential food safety problem. Consumer and government demands for pathogen reduction are increasingly stringent.

Consumer interest in purchasing irradiated food has steadily increased, specially when benefits of irradiation are mentioned. The extensive media coverage following US Food and Drug Administration approval of red meat may have been responsible for increased interest in irradiation used to destroy pathogenic microorganism. Almost 80% of US consumers indicated they would buy products labeled, "irradiated to destroy harmful bacteria," with interest highest in irradiated poultry, pork and ground beef. Consumers have purchased irradiated food whenever it has been available. In recent years over 600,000 pounds of labeled irradiated fruit from Hawaii have sold in the Midwest and California. In market tests up to 80% of Kansas consumers selected irradiated poultry when priced the same or less than non-irradiated poultry and about 20% when the irradiated product was priced at a 20% premium. Currently acceptance of irradiated food is greatest in upscale markets among a more educated public.

A marketing program in Florida featuring endorsement by public health officials has established a demand for irradiated poultry in independent supermarkets. The food service industry has also expressed increased interest in irradiated poultry. Consumers want information on the purpose of irradiation, such as the destruction of harmful bacteria, the safety of irradiated food, the effect of irradiation on nutritional value, and the opinion of the health community. The preferred method to receive information is newspapers and government flyers. Benefits drive acceptance. Most consumers want chicken free of *Salmonella* and *Campylobacter* and meat free of *E. coli* O157:H7. High quality tropical fruit is also appealing to the consumer and the food service industry. These demands can be met through irradiation.

INTRODUCTION

Consumers expect food to be safe, yet press coverage of foodborne illness has increased consumer awareness of the potential severe consequences which can result from infection of *E. coli* O157:H7 and other pathogens. Consumers respond to benefits. Irradiation offers advantages not obtainable by other processing methods. More United States consumers have become aware of these advantages and have heard widespread endorsement by health authorities. An increasing percentage of consumers have indicated interest in purchasing foods irradiated to destroy harmful bacteria. Since the mid 1990s people have purchased fruits irradiated for quarantine control and increased shelf life. Although irradiated foods are only available in very limited regions of the country, plans are proceeding for wide scale marketing of irradiated food in at least one state.

CONSUMER KNOWLEDGE AND CONCERN

Media coverage of illness from *E. coli* O157:H7, *Salmonella* sp., and other pathogens has increased consumer concern about microbiological hazards in food. More people volunteer concerns about microbiological hazards than any other potential food safety problem. Volunteered concern about microbiological safety increased from 36% in 1992 to 69% in 1997 [1, 2].

More consumers believe that food safety problems are likely to occur at the food processor and manufacturers level than any other portion of the food chain. The home is considered a likely problem by only 15% of consumers [2]. Even though increased educational efforts are directed toward consumer safe food handling, people make mistakes and consequences are serious [3-5]. Whether it is appropriate or not, the press often reports serious illness as being due to contaminated food. The responsibility, therefore, is with the manufacturer/processor.

Government demands for pathogen reduction are increasingly stringent when *E coli* O157:H7 is classified as a food contaminant. When found in a lot of ground beef, the manufacturer must recall the product. Notice of recalls appear frequently.

Consumer demands for food safety are increasingly strict. Nancy Donley, Safe Tables our Priority, was quoted in 1998 as saying, "We can not let the safety of our children rest with cafeteria or fast food service worker. We can not protect our kitchen from every single microbe. We must get it out of the food system"[6]. She is calling for a step to destroy pathogens. The only method today that can achieve this objective is irradiation.

Consumer Response to Irradiation

Attitude studies indicate people will purchase irradiated food if the benefits are appealing. In 1996, almost 60% of consumers in a nationwide survey indicated they would buy irradiated foods with longer shelf life and almost 70% would buy food irradiated to destroy harmful bacteria [1]. Sixty percent of consumers indicated they will pay ten cents more per pound for ground beef irradiated to destroy harmful bacteria [7].

When consumers receive information on irradiation, interest in purchasing irradiated food increases. In an Indiana study conducted in 1993, between 60% and 70% of consumers were willing to purchase irradiated food prior to receiving additional information about the process. After exposure to an educational program which included an 8 minute video tape, *The Future of Food Preservation, Food Irradiation,* describing the irradiation process and the effect of irradiation on food safety and wholesomeness, increase in purchase increased to over 80% [8].

The effect of increased information provided by this video tape was examined in communities in California and Indiana. Participants saw *The Future of Food Preservation, Food Irradiation* and were given the opportunity to express concern or ask questions. Those consumers indicating concern about irradiated foods decreased from 19% to 11%. Confidence in the safety of irradiated food increased with those likely or very likely to try irradiated food increasing from 57% to 83% [(9)].

The FDA approval of irradiation of red meat was featured on television and in newspapers and magazines. This extensive coverage may have been responsible for increased consumer acceptance of irradiation used to destroy pathogenic microorganism. In a nationwide study conducted in March 1998 almost 80% of consumers surveyed indicated they would buy products labeled, "irradiated to destroy harmful bacteria" [(10)]. Sixty-seven percent of consumers said it was "appropriate" to irradiated poultry, with pork and ground beef seen as "appropriate" by slightly fewer consumers. Over 60% felt irradiation was appropriate at a fast food restaurant with almost 50% considering it appropriate at the grocery store deli or sit-down restaurant.

Consumers believe irradiation's primary advantage is the destruction of harmful bacteria with almost 80% indicating that as a reason to buy irradiated products. Consumers do not expect irradiation to replace safe food handling. In the 1998 survey, 91% of consumers responded that safe food handling is still important [(10)].

While almost half of consumers interviewed accepted the term, irradiation, cold pasteurization was preferred by 55% [(10)]. This is consistent with focus group studies completed in early 1998 in which consumers indicated they were familiar with pasteurization and could understand the benefits of the irradiation process better if it was paired with this term.

Knowledge of food irradiation and interest in the process continues to be highest among those with higher levels of formal education [(11, 12)].

MARKET PLACE EXPERIENCE

Consumers have purchased irradiated food when-ever it has been available. People have purchased irradiated fruit in Midwest markets since 1992 [(13, 14)]. Since 1995 about 200,000 pounds per year of irradiated fruits from Hawaii including papaya, atemoya, rambutan, lychee and starfruit have sold in the Midwest and California [(15)].

Consumers in Kansas have purchased labeled irradiated poultry in market tests since 1995. The irradiated poultry captured 63% of the market share when priced 10% less than store brand, and 47 % when priced equally. After reading a brief description of irradiation, 80% selected irradiated poultry when it priced the same as the non-irradiated house brand [(11)].

A marketing program in Florida featuring a newspaper advertisement in which the county public health officer endorses the safety of irradiated food, a radio advertisement, and a bill board stating, "PUBLIC HEALTH NOTICE, Pasteurization, Safer Milk, Irradiation, Safer Meat" has established a demand for irradiated poultry in independent supermarkets. The food service industry has also expressed increased interest in irradiated poultry [(16)].

While the US Food and Drug Administration has approved the irradiation of meat, the Department of Agriculture which regulates meat and poultry has not issued regulations for irradiation and labeling. It is therefore, not legal to irradiate beef at this time. Nevertheless, companies which manufacture over 70% of the ground beef sold in the country have contracted with Titan Corporation to use e-beam facilities should they choose to irradiated ground beef. Irradiation of poultry has been approved by FDA and USDA since the mid 1990s. On September 21, 1999, a major poultry processor, Tyson, announced plans to test market poultry irradiated using the Surebeam Titan technology.

CONSUMER INFORMATION

The preferred method to receive information is newspapers and government flyers. Consumers want information on the purpose of irradiation, such as the destruction of harmful bacteria, the safety of irradiated food, the effect of irradiation on nutritional value, and the opinion of the health community. They place the greatest trust in information provided by health professionals [10].

In June of 1999, public health officials in Minnesota held a two day symposium for health professionals reviewing food borne illness and the role of irradiation to increase food safety. The retail food industry was present, both as speakers and audience. A major supermarket chain has announced intention to market irradiated products. The public health web page includes a section on food irradiation.

http://www.health.state.mn.us/disv/dpc/food/foodsafe.htm

CONCLUSIONS

Currently acceptance of irradiated foods is greatest in upscale markets. People are already buying high quality irradiated fruit. Purchase of irradiated meat, poultry and other items will increase with increased consumer education and with public endorsements by health professionals.

Although some consumers will never accept irradiated food, most consumers would like chicken free of *Salmonella* and *Campylobacter* and meat free of *E. coli* O157:H7. Consumer demand for pathogen free products can be met through irradiation.

REFERENCES

1. Abt Associates Inc. 1996. "Trends in the United States, consumer attitudes and the supermarket." Food Marketing Institute. Washington DC. 91.
2. Abt Associates Inc. 1998. "Trends in the United States, consumer attitudes and the supermarket." Food Marketing Institute. Washington DC. 88.
3. Bruhn C. and Schutz H. 1999. "Consumer food safety knowledge and practices," Journal of Food Safety, 19:73-87.
4. Daniels R. 1998. "Home food safety," Food Technology, 52:54-56.
5. Williamson D., Gravani R. and Lawless H. 1992. "Correlating food safety knowledge with home food-preparation practices," Food Technology, 46(5): 94-100.
6. Donley N. 1998. Proceedings of Conference for Food Protection, Boston, MA, April 25, 7-8.

7. Gallup Organization. 1993. "Irradiation: consumers attitudes," American Meat Institute,
8. Pohlman A., Wood O. and Mason A. 1994. "Influence of audiovisuals and food samples on consumer acceptance of food irradiation," Food Technology, 48(12): 46-49.
9. Bruhn C.M. and Mason A. 1996. "Science and Society: A public information program on food innovations." Washington, D.C.
10. American Meat Institute. 1998. "New consumer research good news for irradiated foods." American Meat Institute. Document #16719. Washington, DC.
11. Fox J. and Olson D. 1998. "Market trials of irradiated chicken," Radiation Physics and Chemistry, 52:63-66.
12. Maciorowski K., Ricke S. and Birkhold S. 1999. "Consumer food safety awareness and acceptance of irradiated raw poultry in three Texas cities," Dairy, Food and Environmental Sanitation, 19(1): 18-28.
13. Marcotte M. 1992. "Irradiated strawberries enter the US market," Food Technology, 46(5): 80.
14. Pszczola D. 1992. "Irradiated produce reaches Midwest market," Food Technology, 46(5): 89.
15. Dietz G. Personal Communication: Isometics. 1999.
16. Everett H. Personal Communication: Food Technology Services. 1999.

Florida: A Case Study in Marketing Irradiated Poultry

C. HUNTER
Vice President, Business Development
Industrial Irradiation, MDS Nordion
447 March Road, Kanata, Ontario, Canada K2K 1X8

Abstract:

In the United States, as in many other countries, selling irradiated food is considered by many to be controversial. While the technical benefits and safety of the process are well understood, the general public are often uninformed and their first reaction may be to avoid irradiated foods because they fear of the word "irradiated" on the label. In addition, certain consumer activist organizations have worked to fuel these fears through the effective use of the news media.

The situation is a significant challenge to marketers of irradiated food. This paper discusses recent experience in Florida (the only place where irradiated chicken is being sold in the United States) and provides insights into the challenges and possible solutions to successfully marketing irradiated food. Topics covered include the interests of various stakeholders, issues in advertising and promotion, consumer reactions, sales to restaurants, and the role that government can play in educating the public and influencing industry.

PRESENTATION

Let me start by telling you about a man with a dream, and I will show you how this dream has lead to the first real mass marketing of irradiated poultry in the world.

Unfortunately, I never had the opportunity to meet Sam Whitney as he passed away late last year. He was intrigued about the many benefits that irradiation could bring to the food industry. But as many people here can attest, he was a real character, and unfortunately, his vision of the future was about ten years too soon.

Not being one to let the grass grow under his feet, he formed a company called Vindicator Inc. The initial purpose of the company was to irradiate grapefruit at this plant 45 miles East of Tampa, Florida. The target market was Japan, and the irradiation was to meet quarantine needs.

And so, he installed this Nordion automatic pallet irradiator which became operational in January 1992. Unfortunately, Sam was never able to sell any irradiated grapefruit to the Japanese. They had the same negative attitude towards irradiated food as in many countries around the world.

So, the next question was "What do we do with this irradiator?" Well, later that same year the US Government permitted the use of irradiation to control pathogens in fresh and frozen poultry. So, Vindicator and Nordion started a joint effort to launch irradiated poultry in Florida.

Since none of the poultry companies were ready to start things themselves, they launched under their own Nation's Pride label. The label was designed to try and capture a sense of high quality and make use of a very "American" theme.

The product was launched at Wyndle's Foodland in Plant City Florida. This was near the factory and was a modern supermarket whose owners were willing to take a chance on this new product.

The product itself was fresh poultry in various configurations, and it was given a good amount of shelf space. There were point of purchase signs and information available to shoppers.

This launch was not without controversy, as both the store and the plant were picketed for a short time by protesters, lead by a very small but vocal group called "Food & Water", which passed after a short while.

At the same time, we had a booth at the Florida State Fair where we gave out free samples of irradiated chicken wings and strawberries.

During the 17 days of the Fair, we gave out 23,000 chicken wings and 5000 strawberries. 95% of the people tried the product and only 16 raised any objections. A very positive result, especially given the relatively low profile of foodborne pathogens at that time.

Now let me take a moment to point out something very interesting. At first the sign on the booth said "Irradiated". After a while they changed it to the word "Zapped". The result was that the public had a much friendlier attitude towards the process. The word "Zapped" did not carry with it the same negative baggage that unfortunately goes with "Irradiated". Because "Zapped" is a more casual and familiar word, it does not start out threatening the customer.

This is the first lesson from the Florida case study: "In any promotion, find a way to make consumers feel at ease with the product, ideally by tying it to a concept with which they are familiar."

Unfortunately, the market was not ready for irradiated food then. Neither were the poultry processors. To understand why, we need to look at their motivations. Everybody has been long aware that poultry is often contaminated with *Salmonella* bacteria, so poultry did not have an image problem relative to other types of meat. They did not perceive that they had a problem, so they did not want to raise any concerns with their customers. Also, irradiation was (and many would say still is) a very controversial technology. It was a very simple decision for them not to take a significant risk when their market has been growing every year.

Everything started to change with this headline in 1997. *E. coli* O157:H7 is now known to most people in North America as "Hamburger Disease". This concern has been increased with ongoing reporting of disease outbreaks, including the *Listeria* outbreak that killed 21 people at the end of 1998. All of this has caused food borne illness to be rated as the number one concern that people in the US have their food supply. Even though these diseases are not linked to poultry, the poultry industry has been caught up in the public concern.

So, let us to return to Florida. By this time, Vindicator Inc. had changed its name to Food Technology Service Inc. They now wanted to take advantage of these changing attitudes to improve their business.

However, they still ran into a problem. When they went to the poultry processors, they were told that they were not interested in selling irradiated food until their customers asked for it. When they went to the supermarkets they were told the same thing. This was a real dilemma for them – they are not big enough company to do mass advertising to convince consumers directly.

Food Technology also tried to approach restaurants. The fast food chains are not interested in leading, because they think the risk is too great, given their high public profile. They will likely move once others start to take the lead. To get smaller restaurants to sell irradiated poultry, Food Technology first had to get Distributors to carry the product. But the distributors would not carry it if there was no market. Another dilemma. Which comes first so to speak, the chicken or the egg?

Food Technology approached these problems from two directions. In the first, they have started working with Colorado Boxed Beef, a $600 million turnover beef processor & packager located near their plant. Colorado Boxed Beef wants to be one of the first on the market to aggressively promote irradiated beef once the US Department of Agriculture finally releases the regulations to allow its sale. The reason this came about is because Colorado Boxed Beef is a small enough that it does not have the same perceived risk as the largest beef producers. But it sees a lot of upside potential in gaining market share by being first.

So my second lesson from this case study is that to be successful you need to work with companies who have the right motivation to make Irradiated Food a success. Other major meat companies have announced that they will be testing irradiated food and will soon follow the aggressive position of Colorado Boxed Beef.

The second of Food Technology's initiatives was to advertise publicly. They did this not on a broad scale, but in the county where they are located, thus making it affordable. This is their first full page newspaper advertisement. Notice how the ad works:

- First, it draws a comparison between pasteurized milk and irradiation. It also uses a picture of some children and emphasizes the benefits for a family. It is targeted at the mother of a household who would normally be buying the food.
- Food Technology also relied heavily on endorsements of independent organizations like the WHO and various American medical organizations. They got the local county health department to respond to consumers questions.
- Finally, the advertisement suggests that consumers should ask their supermarket for irradiated food. This was to try to address the dilemma

Food Technology also ran some radio commercials. But perhaps the advertisements with the most impact were these billboards. The message is very clear. The hamburger reminds consumers of *E. coli* O157:H7. The connection with pasteurized milk ties the irradiation process to a familiar and comfortable process, while targeting mothers with children. Finally, the heading "Public Health Notice" makes it seem like the advertisement was sponsored by the Florida Department of Health. And in fact, Food Technology has worked with the Health Department to get their active support.

They used this in their next newspaper advertisement, which followed the same theme as the billboards. The lower half of the advertisement is a letter from Dr. Richard Hunter, the Deputy State Health Officer for Florida. It talks about the growing incidence of food borne illness, referring specifically to *E. coli* and the higher susceptibility of children and the elderly. The letter also covers the safety of the process, the independent organizations that endorse it, and the comparison to pasteurized milk. He concludes by saying: "I hope you will purchase irradiated chicken and ground beef as they become available. This combination will afford your family maximum protection against foodborne illness." Dr. Hunter sent similar letters to the heads of all the large supermarkets in Florida and to the chain restaurants, urging them to stock irradiated food.

This brings us to the third lesson from this case study. You can increase your credibility and gain more influence if you get public health authorities to actively support radiation processing. This assumes that they have a good reputation to bring. Another benefit of their support is the free publicity that they can bring to the process.

So, what has been the result of all this? The first is that the same Wyndle's Foodland and one other store have started carrying irradiated chicken again. But this time it is individually quick frozen product packed in 2½ and 4 pound bags.

Food Technology has also had success at expanding into the food service market. They now have five different distributors handling the product. This includes chicken produced under the Harvestland label shown here

Note that one of the things that makes it somewhat easier to get restaurants to try irradiated food is that there is no need to identify it on the menu.

In conclusion, we believe that Florida will turn out to be a good location for the sale of irradiated poultry and that this case study shows some valuable approaches that can be used to make this process a commercial success. And after all, there is not much point in going through all the effort in developing such a process if no one ever uses it. So I think that Sam Whitney would be pleased to see what is starting to come from his vision for a business in Southern Florida.

Growing Impact of Irradiation on Global Production of and Trade in Spices

M. I. EISS
Technical Activities Director
American Spice Trade Association

ABSTRACT

Spice consumption is increasing world-wide. In 1998, the United States per capita consumption of spices and herbs increased to 3.6 pounds or 1.64 kilos per person. Total consumption (imports and domestic production, minus exports) reached a record 968,000,000 pounds (440,000,000 kilos). U.S. 1998 imports totaled 292,700,000 kilos. Major spice supplying countries include Canada, India, Indonesia, China, Mexico, Guatemala, and Turkey. If the current trend continues, the U.S. can break the billion pound mark by the beginning of the new millennium.

Along with increased consumption has come heightened concern for spice cleanliness. The health conscious consumer, tougher environmental regulations, intense competition in the area of manufactured foods, and a changing global economic structure have led to major changes in the spice industry and the food industry in general. The trend in spice usage has gone from a consumer base to an industrial base. Over 60% of the United States spice usage is industrial/food service (catering). Retail sales account for less than 40%. The trend in Europe is similar. In addition, the clear trend in the industrialized countries is one of increased legislation and inspection with regard to spice sanitation and contamination. In the U.S, there is a major emphasis on food safety and the reduction of microorganisms in imported foods. There are at least two bills pending in the United States Congress to have stricter controls on imported products. Spices are present in almost every manufactured food in the United States. Many of these foods are prone to spoilage from microbiological contamination. The impact of spices on processed foods is great.

The American Spice Trade Association has been working with major spice and seasoning companies and with growers in source countries to improve spice quality and cleanliness. New technologies and practices are evolving regularly in both traditional source countries and most noticeably, in some of the newer producing areas to help ensure better and cleaner products. Since spices are an agricultural commodity, and it is not possible to grow a perfect spice, some contamination often occurs. After many years of testing, irradiation has proven to be an important means of reducing microbial contamination in spices. Commercial ionizing radiation treatment of spices has been in place since the mid- 1980s. In the early years, production was low. However, since 1991, the sales of irradiated spices, herbs, and vegetable seasonings have increased steadily. Approximately 34 million kilos of product were irradiated in North America in 1997. The poundage for irradiated product for 1998 is estimated to be about 38.6 million kilos. Many years of testing have shown that spices, herbs, and vegetable seasonings respond well to irradiation doses up to 30 kGy. Levels of microorganisms are greatly reduced, and the spice products retain their sensory qualities. Spices treated with irradiation can be shipped, stored or eaten immediately after processing. The irradiation process kills harmful bacteria while providing consumers with a safe wholesome product.

The term "spice" originated from the Latin "species aromatacea" meaning fruits of the earth. The term was later shortened to "species" meaning a commodity of special value or distinction. Of course, many spices such as black pepper have had special value throughout their long history. Black pepper was so valuable during the middle ages that pepper berries were often used to pay rents. The American Spice Trade Association (ASTA) defines spices as "any dried plant product used primarily for seasoning purposes".[1] Although organizations such as ASTA, the U.S. Food and Drug Administration[2], and the International Standards Organization/Geneva[3] all have different official lists for spices, they generally include tropical aromatics such as pepper, cinnamon, and cloves; leafy herbs of the temperate zone, oregano, basil, sage, etc; spice seeds such as sesame, mustard, caraway, poppy seed, and dill; and the dehydrated vegetables onion and garlic. All of these products play an important part in flavoring foods and have been a very important part of international trade for many centuries.

In modern times, as transportation has become easier and the world smaller, spice consumption has increased dramatically. In 1998, the United States per capita consumption of spices and herbs was 3.6 pounds/1.64 kilograms per person. Total consumption (imports and domestic production, minus exports) reached a record 968,000,000 pounds/440,000,000 kilos.[4] If the current trend continues, the U.S. can break the billion pound mark by the beginning of the new century.

Along with increased consumption has come heightened concern for spice cleanliness. In the spice importing countries, the health conscious consumer, tougher environmental regulations, intense competition in the area of manufactured foods, and a changing global economic structure have led to major changes in the spice industry and the food industry in general.

The trend in spice usage has gone from a consumer base to an industrial base. Over 60% of the United States spice usage is industrial/food service (catering). Retail sales account for less than 40%. This growth has been driven by changes in eating patterns including a strong consumer demand for highly seasoned ethnic foods at home and from manufacturers, and greatly increased consumption of away from home meals, including fast foods. Contributing to this change, the U.S. Asian population, notable consumers of highly seasoned foods, doubled in the ten years between 1982 and 1992. It is estimated that Americans eat more than one-half of their meals in public eating establishments. According to ASTA statistics,[4] Americans are using 50% more spice today than they were a decade ago. Interestingly enough, the hottest spices, black and white pepper, red pepper, mustard seed, and ginger, have increased over 70% in the past 20 years. The largest gains have been made by the hottest spice, red pepper. The trend is predicted to continue.

The trend in other major importing countries is similar. In Europe, the spice market, which is estimated to be over $U.S. 2 billion, is about 44% retail 41% industrial, and 15% food service/catering. Therefore, the industrial/catering segment of the European market is over 56%. Germany is the principal European consumption market. Their largest spice trading partner is Indonesia from whom they purchase black and white pepper, nutmegs, and cinnamon. Other major European importers include France, the UK, Netherlands and Scandinavia. The third largest spice importer, and the largest in Asia, is Japan which had a startling growth of spice imports of 75% in the five years between 1989 and 1993. Other growth areas that have shown upward trends in spice consumption are the Middle East, and Latin American countries, such as Mexico, who are significant importers of cinnamon and cassia.

Spices are truly a global business. The current top supplying nations to the U.S. are Canada, India (which grows the largest number of spices and is also the world's largest consumer), Indonesia, China, Mexico, Guatemala, and Turkey (Table 1). The top twelve spices imported into the United States are listed in Table 2. The United States produces about 40% of its own consumption. The domestic crops include dehydrated onion and garlic, paprika, chili peppers, and mustard seed. U.S. production was down for 1998 principally because of the very poor garlic crop that was a 50% loss due to weather (El Nino) and disease. Along with increased consumption has come heightened concern for spice cleanliness. The clear trend in the industrialized countries is one of increased legislation and inspection with regard to spice sanitation and contamination. In the U.S., there is a major emphasis on food safety including the reduction of microorganisms in imported foods. There are at least two bills pending in the United States Congress to apply stricter controls on imported products. These bills, if passed, will give the Food and Drug Administration more authority to reject imports. In addition to the Congressional bills, there is a government task force from President Clinton's Council on Food Safety[5] that is looking at various issues.

Although microbial contamination is the primary focus, chemical contamination and pesticides will also be addressed. Not only is the concern about food safety and cleanliness of food increasing in the U.S., it is also happening in the other major spice importing countries of the world. In Europe, the head of the European Commission has called for the formation of an independent agency to help restore the public's faith in the quality and safety of foods.[6] The new organization is to be modeled after the U.S. Food and Drug Administration. The UK is currently discussing a Food Standards bill which aims to protect public health.

In earlier years, we accepted the fact that improvements in the sanitation level of imports were difficult to achieve because of the "cottage industry" nature of the spice business coupled with the long supply chains. Given the rather poor quality coming into the importing countries, improvement of raw material cleanliness was accomplished by expanding cleaning and processing capabilities in America, Europe and Japan. And, while this was not ideal, it was possible under the then current conditions. However the supply chain has changed dramatically over the past 15-20 years. The numbers of brokers in both exporting and importing countries are being reduced. Large multinational companies, and even some smaller ones, are going directly to the source. For example, in Germany in 1970, there were about 75 spice grinders and approximately 20 importers and brokers. By the mid- 1990's, there were about 15 importers and brokers supplying product to 50 grinders and the consolidation trend is expected to continue. In the UK the number of brokers/importers has decreased from 15 to approximately 5, and the number of spice processors has decreased from 75 to less than 50. The shrinking supply chain is the same in the exporting countries. India, which had hundreds of brokers 25 years ago, has seen the numbers change dramatically.

The industrial customer is now the focus. International catering/fast food giants such as McDonald's, KFC, Burger King, and Pizza Hut have tremendous buying power with their central purchasing departments. Mergers in the food industry have increased the strengths of multinational companies. These trends have led to tightened specifications, because industrial and catering firms cannot readily accommodate large variation in their raw materials if their food processing lines, and businesses, are to run efficiently and products with tight industrial specifications are to be manufactured. These trends have also led to increased inspection and regulation by such bodies as the FDA and the European Union health authorities. In the U.S., FDA inspections have increased 25% and are expected to increase more as pathogens become a major focus. Microbial reduction is a major thrust for authorities in such places as the U.S. and Europe. In a product recall in Germany in 1993 of paprika potato chips contaminated with salmonella, the recall cost the parties involved an estimated minimum of US$10,000,000. The industrial concern cannot afford, and will not tolerate, such incidents. In addition to the actual monetary cost, the good name of their food products and business is at risk.

Spices are present in almost every manufactured food. Many of these foods are prone to spoilage from microbiological contamination. While spices are used in food products in small amounts, generally less than 2%, the impact of spices on processed foods is great. The American Spice Trade Association has been working with major spice and seasoning companies and with growers in source countries world-wide to improve spice quality and cleanliness. There are, of course, microorganisms that are indigenous to the soil, but other sources of contamination include dust, insects, and fecal materials from birds, rodents, and other animals. Fungal growth may occur at several stages of spice processing — prior to drying, drying, storage and shipping. New technologies and practices are evolving regularly in both traditional source countries and most noticeably, in some of the newer producing areas to help ensure better and cleaner products. And, while progress is being made, there are still problem areas. An all inclusive approach to cleanliness is needed. There should be prevention of contamination from planting through the final phase of post harvest processing and shipping. To accomplish this, action must be initiated across a broad spectrum, including the improvement of growing, harvesting, dehydrating and raw material storage, and implementation of training programs at all levels in the production chain.

Since spices are an agricultural commodity, and it is not possible to grow a perfect spice, some contamination often occurs. Spices and herbs commonly harbor large numbers of bacteria and fungi including potential spoilage organisms. In general, roots, berries, and herbs carry a greater microbiological load than the bark and seed items. Of all the spices, black pepper typically has the highest aerobic plate counts, usually in excess of 10^6-10^8 cfu/g. Paprika, celery seed, coriander, turmeric, thyme, basil and others can also have plate counts in the millions per gram. The most common bacteria are the aerobic spore-formers, such as Bacillus, clostridium, and salmonella species.[7,8,9] Common molds include some of the Aspergillus, penicillium and Rhizopus species. Nonspore forming bacteria, including indicator organisms and those of a potentially more serious nature, may also be present. Coliforms are often detected at import along with *E.coli* .

Although spices are employed as minor ingredients in food products, they nevertheless contribute a significant population of microorganisms, that find an ideal culturing medium in many semi-processed and even processed foods. As a result they can cause food spoilage. The food industry has been concerned for many decades with finding methods to decontaminate spices. Today, three major methods are in use, gas (ethylene oxide and propylene oxide), irradiation, and steam. Ethylene oxide gas was the sterilant of choice for many years. However, today it is banned in the European Union and other countries. Ethylene oxide is not effective on some products due to the fact it requires about 7.5-8% moisture level for use. Steam is a natural treatment and has no regulatory barriers. It is effective for use on berries and seeds. However, steam degrades the light-weight leafy herbs, and ground products are difficult and sometimes impossible to handle in the steam systems.

Irradiation is a decontamination treatment that has been tested for many years. Sterilization by ionizing radiation has a major advantage in that it is a cold treatment. Irradiation dosage as high as 50 kGy will only increase the temperature of food by 12 degree C. Therefore, there is little danger of the loss of volatile components of the spice. Other advantages of irradiation include: ability to irradiate in bulk packages, capability to process salt blends, and no residues.

In the early 1980s, a major American spice company initiated a major testing program for the irradiation of over twenty spices. The test results were all favorable and were used in petitions for spice clearances in Canada, the United States, and other countries.[10,11]

All test work was done with gamma irradiation from cobalt 60 because of its penetration power. Testing was done in various sizes of commercial packages — drums, boxes, bags, jars, and tins (50-2000 lb lots). A commercial scale-up sample of 40,000 lbs was also irradiated. Testing covered microbiology, chemistry, sensory/organoleptic analysis, food application studies, and wholesomeness studies.

Microbiology[12]

Approximately 20 spices were analyzed for microbial contamination. Results are listed in Table 3. Black pepper samples were the most contaminated, but they also cleaned easily. Onion powder was consistently the most difficult product to clean. All spices were irradiated at a 10 kGy average dose, except for the capsicums, paprika and red pepper, which were irradiated at a lower dose of about 6-7 kGy. The average dose was commonly used in the 1980s.

In the commercial scale-up, 40,000 pounds of various onion powders in 800/50 lb boxes were processed. The onion was high in bacteria with an average Standard Plate Count of about 7,000,000. Reduction of microorganisms was approximately 98%. Over the years irradiation techniques have been perfected and improved. A table was recently developed to give suggested irradiation doses ranges for various spices, herbs, and vegetable seasonings. The table is published in ASTM Standard Guide F 1885-98[13]. Average dose is no longer used. A maximum and minimum dose are listed. The dose ranges are only suggestions. If initial micro analysis indicates a higher than normal plate count, higher minimum doses may be required. The absorbed dose range for a given product depends on the spice density, the type and number of microorganisms in the unprocessed spice, the radiation sensitivity of the organisms present, and the residual number of non-pathogenic microorganisms considered acceptable by the customer. Different spice uses allow different amounts of microorganisms in the end product. For example, seasonings and spices for meat products generally require low micro counts. Meats provide an ideal culturing medium for microorganism growth.

However, if the end product will be retorted, cooked at a high temperature, higher levels of microorganisms can be tolerated. Additionally, some countries have specific microbiological specifications for all imported spices and seasonings. These specifications must be met.

Chemistry[12]

Dry products such as spices are less affected chemically by irradiation than high moisture foods such as beef and whole fruits. Generally, impractically high radiation doses would be required to form significant levels of radiolytic products in spice products. They are very stable. Extensive chemical analysis of onion powders revealed very few chemical changes. Constituents investigated were volatiles, amino acids, sugars, starch, carbohydrates, protein, and fat.

Most spices are used for flavoring foods. However, paprika is used for its color. Shelf-life studies were carried out to monitor the effect of irradiation of the coloring pigment, capsanthin, in paprika. Six and twelve month storage at temperatures of 40^0, 75^0, 90^0 F showed no color differences.

Sensory/organoleptic Analysis[12]

Spices are used to flavor foods both industrially and in the home. The amount of spice used in a product depends on the flavor strength desired by the consumer. While the objective in irradiating spices is reduction of microbial levels, it is of prime importance that flavor not be adversely affected by the process.

Extensive sensory/organoleptic analyses were done on the irradiated products using trained panelists with a broad base of experience of different food products. The following standard procedures were used: Bench-top evaluations, triangle tests, paired comparison (for black pepper to minimize sensory fatigue), special red pepper analysis[14], and descriptive analysis. Few differences were found in any of the spices tested.

As a further check of the taste properties of the irradiated spices, all examples were evaluated in three types of food applications, hot, hot/cold, and cold. For example: celery seed in potato salad, herbs in soup, pepper on eggs, etc. No major differences were found.

Wholesomeness

Government approvals for food additives, and unfortunately in the U.S. and other countries irradiation is classed as a food additive, require wholesomeness testing, mutagencity testing and/or feeding studies. Mutagenicity tests conducted on dehydrated onion powders were: the Ames test, *in vivo* sister chromatid exchange, and drosophila melanogaster. Results of these experiments were published. No adverse effects were found in any of the tests.[15,16,17]

Animal feeding tests conducted with pungent materials, such as many of the spices and herbs, are extremely difficult, and in many cases impractical, particularly at the high levels of solids intake required by regulations. In a number of cases, in both short and long term feeding studies, investigators found various toxic effects due to components in the spices rather than due to irradiation.

All of the work described thus far was done with gamma irradiation from cobalt 60 Recently, work has been done to determine the practicality of using low power self-shielded electron beam equipment to sterilize spices at the end of a production line. The goal was to demonstrate the feasibility of using high dose rate capabilities of a curtain type electron processor for the treatment of thin fluidized beds to obtain high throughput. Seeds, berries, leafy herbs, and powders were tested. Work was experimental and the equipment small scale pilot. The product handling utilized standard pneumatic transport techniques, including vibratory feed, filter bag recovery and regenerative blower transport of the product. The pilot processor was a 30 cm wide, 250 kV Electrocurtain unit. Results were very good for seeds, berries, and leafy herbs. During the first experiments, powders tended to burn under the beam. Modifications were made to the system in order to be able to handle powders. The microbial reduction is excellent in this system (Table 4); the taste of the product remains good. Much work remains to be done to commercialize this type of system, but the concept is interesting for processing large amounts of raw materials.

Over the years it has been proven that irradiation, gamma or electron beam, is an excellent process for decontaminating spices. However, irradiation is not a substitute for good sanitation. If the product is not maintained in intact sealed packaging, and in a clean warehouse, protected from animals, birds, insects, and rodents, all of the work is in vain. Product must be stored in a closed building neatly off the floor and away from the walls, with the proper rodent and insect traps. If the cleaned product is stored in a dirty, improperly maintained warehouse, rodent and insect infested, it will probably become recontaminated. Good packaging and storage conditions are critical.

Commercialization

Overall, the commercial development of spice irradiation has been slow. This has been frustrating but not surprising. Consumers world-wide are, and always have been, slow to embrace new food technologies. The history of canning, which is one of the greatest food processing inventions ever made, tells us that is was over 50 years before consumers accepted canning as a safe food process. Pasteurization also had problems with acceptance. Spice irradiation has only been a viable commercial process for 10-15 years.

Recent events indicate that food companies and the public are changing their attitude toward irradiation processing. In the United States, the Food Irradiation Coalition has been formed. The Coalition, under the auspices of the National Food Processors Association, is composed of a number of the largest food trade associations in America. These include the Grocery Manufacturers Association, American Meat Institute, the National Restaurant Association, ASTA, Food Marketing Institute, the Packaging Society, and others. The associations are being urged by their members, the food manufacturers and restaurants, to petition for additional uses of irradiations and for changes in regulations.

International Trade

There are several issues that further impact on international trade in irradiated spices and herbs: Labeling/dose allowed, detection of irradiated product, and consumer acceptance.

Labeling/Dose allowed — In the United States, spices may be irradiated up to a dose of 30 kGy.[18] The U.S. has a regulatory situation in which the FDA regulates labeling of almost all food products except meats, which are regulated by the Department of Agriculture. Under FDA regulations all bulk ingredients and retail products must be labeled. Irradiated ingredients shipped into the country must also be labeled. However, currently, the labeling requirements apply only to foods that have been irradiated, not to foods that contain irradiated ingredients. The U.S. Department of Agriculture, which has jurisdiction over meat products, is currently formulating meat regulations and is proposing that all meat ingredients be labeled.

The labeling of minor ingredients and second generation products produced with irradiated ingredients is a complicated and controversial issue. The views of other governments vary from the U.S., which requires no labeling, to the new EU regulation[19] that will require all irradiated ingredients to be labeled. The labeling issue is, to some extent, a barrier to trade. U.S. processed food manufacturers, right or wrong, do not want to label ingredients. One reason is the very high cost associated with changing labels on thousands of products.

As noted above, the U.S. standard allows spices to be irradiated to a maximum of 30 kGy. When most other countries granted clearances for irradiated spices, the older 10 kGy average dose standard was adopted. The 30 kGy maximum dose is a more practical limit. There needs to be harmonization of world-wide allowed dosage.

Detection Methods for Irradiated Foods[20,21]

Laboratory tests, that can identify irradiated products, would be advantageous to have to enforce labeling rules for the countries interested. Many governments have spent years looking for test methods. In general, no reliable routine tests have been developed for simple identification of all irradiated foods. Some tests show promising results for specific products. For spices, the best method currently is thermoluminescence. The thermoluminescence test is based on the principle that electrons that are transferred to an excited state by ionizing radiation, and trapped in the solid matrix, return, with the emission of light, to a ground state when irradiated solid samples are heated. By measuring the light intensity as a function of temperature, the so-called "glow curve" is obtained. Thermoluminescence intensities are different from one spice to another, and individual spices give different dose-response curves. Studies have demonstrated that the thermo signal is basically due to contaminating minerals (silicate materials) present in variable amounts in many spices and herbs. Photostimulated luminescense is sometimes used as a rapid screening technique in combination with thermoluminescence for detection of irradiated products.

Consumer Acceptance

The major question in the industrialized countries is "Will consumers accept irradiated foods?" 15 years ago there was little understanding of the irradiation process by the public. The situation has changed. Press articles are changing. In place of dramatic headlines discussing irradiation, most recent news stories are favorable to irradiation. Consumer education is still needed to help the public, but studies show that, when educated about irradiation, many consumers are willing to try irradiated foods. Every test market of irradiated foods in the U.S. has been successful.[22,23] Commercial ionizing radiation treatment of spices has been in place since the mid-1980s. In the early years, production was low. However, since 1991, the sales of irradiated spices, herbs, and vegetable seasonings have increased steadily. Approximately 75 million pounds of product were irradiated in North America in 1997. The poundage for irradiated spices, herbs, and vegetable seasonings for 1998 is estimated to be about 85 million pounds/38.6 million kilos (Figure 2). Irradiated spices are a small but growing percentage of the total world spice trade. The amounts are growing as new irradiation facilities open around the world. Serious problems such as *E.coli* and *Listeria* outbreaks, have led both food processors and consumers to give serious consideration to alternate methods of food processing — including irradiation. From China to Brazil to the Netherlands, spice products are being irradiated.

Many years of testing have shown that spices, herbs, and vegetable seasonings respond well to irradiation doses up to 30 kGy. Levels of microorganisms are greatly reduced, and the spice products retain their sensory qualities. Spices treated with irradiation can be shipped, stored or eaten immediately after processing. The irradiation process is technologically and economically feasible while providing consumers around the world with a safe wholesome product.

Historically, the spice market has be governed to a large extent by the simple economic factors of supply and demand. Today, QUALITY and CLEANLINESS play a very important part in the spice trade. If exporting countries do not upgrade and invest in the future, many will be unable to supply the industrialized importing countries. In order to compete in the spice trade in the coming millennium, supplying countries must plan what action is necessary to maintain the flow of high quality, clean and safe spice products.

References

1. American Spice Trade Association. 1992. "Spice List". Englewood Cliffs, NJ, ASTA
2. U.S. Food and Drug Administration. 1993, "Flavoring Agents and Related Substances". Title 21 Code of Federal Regulations, Part 172.510, Subpart F.
3. International Organization for Standardization. 1995. "Spices and Condiments -Botanical Nomenclature. ISO Standard 676
4. American Spice Trade Association. 1998. "The ASTA Report Spice Statistics 1998 - Imports, United States Production, Current Trends". Englewood Cliffs, NJ, ASTA
5. Mermelstein, N.H., editor. 1999. "Washington News - President Announces Measures to restrict Importation of Unsafe Foods". Food Technology, 53(8):28
6. 1999. "Prodi Wants European Food Standards Agency". The Grocer, July 31, 1999, p. 11
7. McKee, L.H. 1995. "Microbial Contamination of Spices and Herbs: a Review". Lebensm. Wiss. U Technol., 28:1-11
8. Kneifel, W. and Berger, E. 1993. "Microbiological Criteria of Random Samples of Spices and Herbs Retailed on the Austrian Market". J. Food Protection, 57:893-901
9. Pafumi, J. 1986. "Assessment of the Microbiological Quality of Spices and Herbs". J. Food Protection, 49:958-963
10. McCormick & Company. 1982. "Petition to Food Directorate, Health Protection Branch, Health and Welfare Canada for Irradiation Processing of Dehydrated Onion Powders". Hunt

Valley, MD, McCormick & Co.

11. McCormick & Company. 1993. "Petition to Food Directorate, Health Protection Branch, Health and Welfare Canada for Irradiation Processing of Spices and Herbs". Hunt Valley, MD, McCormick & Co.
12. Eiss, M.I.. 1984. "Irradiation of Spices and Herbs". Food Technol Australia, 36:362
13. American Society for Testing and Materials. 1998. "Standard Guide for Irradiation of Dried Spices, Herbs, and Vegetable Seasonings to Control Pathogens and Other Microorganisms". ASTM Standard F 1885-98
14. American Society for Testing and Materials. 1990. "Standard Test Method for Sensory Evaluation of oleoresin Capsicum". ASTM Standard E 1396-90
15. Munzer, R. and Renner, H.W. 1981. "Mutagenicity Testing of Irradiated Onion Powder". J. Food Science, 46:1269-1275
16. Renner, H.W. 1980. "In Vivo Mutagenicity Testing of Irradiated Onion Powder Using the Sister Chromatid Exchange (SCE)" Institute for Biochemistry, Federal Research Centre for Nutrition, Karlsruhe
17. Mittler, S. and Eiss, M.I. 1981. "Failure of Irradiated Onion to Induce Sex-Linked Recessive Lethal Mutations in Drosophila Melanogaster" Mutation Research Letters: 1981
18. U.S. Food and Drug Administration. 1995. "Ionizing Radiation for the Treatment of Food" Title 21 Code of Federal Regulations, Part 179.26
19. European Commission. 1999. "On the Approximation of the Laws of the Member States Concerning Foodsand Food Ingredients Treated with Ionising Radiation". Directive 1999/2/EC of the European Parliament and of the Council of 22 February 1999
20. Schreiber, G.A., et al. 1993. "Thermoluminescence Analysis to Detect Irradiated Spices, Herbs and Spice-and-Herb Mixtures --- An Intercomparison Study". SozEp Hefte, Institut fur Sozialmedizin Epidemiologie des Bundesgesundheitsamtes.
21. Sanderson, D.C.W., et al. 1998. "Establishing Luminescence Methods to Detect Irradiated Foods". Food Science and Technology Today, 12(2)97-102
22. Bruhn, C. M. 1995. "Consumer and Market Response to Irradiated Food" J. Food Protection, 58, p. 175
23. Bruhn, C.M. 1997. "Consumer Acceptance of Irradiated Food: Theory and Reality". 10th International Meeting on Radiation Processing, Anaheim, CA., May 11-16, 1997

Table 1: MAJOR UNITED STATES SPICE SUPPLYING NATIONS

NATION	PRODUCTS
United States	Dehydrated onion and garlic, paprika, chili peppers, mustard seed
Canada	Mustard seed, coriander seed, caraway seed
India	Black pepper, sesame seed, red pepper, cumin seed, turmeric, celery seed, fennel
Indonesia	Black pepper, cinnamon, white pepper, nutmeg, vanilla
China	Garlic, red pepper, ginger
Mexico	Sesame seed, red pepper, oregano
Guatemala	Sesame seed, cardamom seed

Table 2: **TOP 12 SPICES 1998**
(By volume)

SPICE	SUPPLYING NATION
Dehydrated onion/garlic	U.S., China
Mustard seed	Canada, U.S.
Sesame seed	Guatemala, Mexico, India
Red pepper	U.S., Mexico, China, India
Black pepper	India, Indonesia, Brazil
Paprika	U.S., Spain, Chile, Israel
Cinnamon	Indonesia, Sri Lanka, China
Cumin seed	India, Turkey, Syria
Oregano	Turkey, Mexico, Greece
Poppy seed	Australia, Netherlands, Turkey
White pepper	Indonesia, Malaysia, China
Fennel seed	Egypt, India, Turkey

Table 3: **MICROBIOLOGICAL DATA OF GAMMA IRRADIATED SPICES**

SPICE	STD. PLATE CT	YEAST	MOLD	COLIFORMS
Allspice control	2,278,000	LT10	0	0
1000 Krad	LT10*	LT10	0	0
Oregano control	1,214,400	40,000	9,000	2600
1000 Krad	LT10	LT10	LT10	0
Blk pepper control	32,000,000	0	0	25
1000 Krad	60	0	0	0
Basil control	3,000,000	30,000	400	11,000
1000 Krad	1,000	LT10	LT10	0
Paprika control	2,200,000	0	0	410
650 Krad	260	0	0	0
Onion control	1,100,000	0	1,000	170
1000 Krad	11,000	0	LT10	0

*LT=less than

Table 4: **MICROBIOLOGICAL ANALYSIS OF ELECTRON BEAM TREATED SPICES**

SAMPLE	STD. PLATE CT. COLIFORMS	YEAST	MOLD	
Ground nutmeg				
Control	710,000	LT10*	10,000	LT3
5 kGy	1,700	LT10	LT10	LT3
10 kGy	100	LT10	LT10	LT3
15 kGy	20	LT10	LT10	LT3
Ground pepper	8,200,000	LT10	120	4
5 kGy	20,000	LT10	LT10	LT3
10 kGy	280	LT10	LT10	LT3
15kGy	90	LT10	LT10	LT3

*LT=less than

Prospects of Irradiation as a Phytosanitary Treatment of Chilean Grapes

T. C. RUBIO
Comisión Chilena de Energía Nuclear

Abstract

Table grapes are one of the most important fruits produced and exported by Chile. During the season 1997-1998 more than 490,000 metric tons of this fruit were produced, The last season's exportation of grapes meant an approximate income of US$ 604 millions. The false grape mite, *Brevipalpus chilensis*, is a species occasionally found on table grapes, lemons and kiwis. Since *B. chilensis* is considered a quarantine pest for the Chilean table grapes' main market (USA), and on the other hand, since the size of this acaride mite (less than 1 mm), the fumigation of this product before the exportation is mandatory. Methyl bromide is the quarantine treatment most used in Chile for this purpose, but it will be phased out during next decade under the term of Montreal Protocol because it has potential side effects on the atmospheric ozone layer. Due to the reason mentioned before it is urgently needed to find a suitable alternative to the use of methyl bromide as a fumigant for Chilean table grapes.

Since some years ago, several potential alternatives are being studied in Chile. At the beginning of 80's, the Chilean private sector, in collaboration with government institutions such as Comision Chilena de Energia Nuclear, started thc studies on the use of ionizing radiation as a quarantine treatment. Sixteen different fresh fruit species, including 3 varieties of grapes were studied, using low doses of ionizing radiation, in order to know a possible effect of the treatment on the commercial quality of these agricultural products. As a result of these studies it was demonstrated that an average dose of 1 kGy do not affect the commercial quality of tables grapes and most of the other fresh fruits.

At the mean time, several studies have also been done from the 80's up to now, in order to know the effect of ionizing radiation on different stages of development of the acaride mite just mentioned. In the last one of these studies, doses of 0.5, 1.0 and 1.5 kGy were used, combined with 24 hours precold storage before irradiation and an additional period of 15 days of storage at 0-2° C and 90% RH (to simulate shipping conditions from Chile to USA). The doses mentioned were applied on eggs, immature stages and adults of *B. chilensis*, maintained on *Ligustrum japonicurn* L. leaves. It was demonstrated that the irradiated acaride mites were unable to continue their cycle and they would not represent a risk to countries importmg these products from Chile.

The establishment of commercial irradiation plants have also been analyzed for operation in the three most important Chilean ports from where table grapes are exported. The irradiation process proved economically feasible as well. At present, the possibility of building small irradiation facilities next to fruit packing is also being considered. The Chilean private sectors and governmental institutions are continuously working together in order to find soon an alternative to the methyl bromide treatment and where ionizing radiation is being seriously analyzed

In summary, low doses of irradiation seem to be a good quarantine treatment for Chilean table grapes from the technological and economical point of view. However, the possibility to apply this technology at commercial scale in the near future will depend not only the national but also the international efforts such as the establishment of an accepted international protocol for the research studies, approval of the technology by government of the importing countries, consumer acceptance and the feasibility of the other alternative treatments.

Introduction.

Chile is a major producer and exporter of fresh fruit and vegetables. Table grapes are the most important product in terms of volume. During the season 1997-1998, Chile exported a total of 67.148.052 boxes of table grapes and the main market is USA (55,5 %). The fumigation with MB is mandatory to the entrance of table grapes to USA, but this fumigant is being phased-out globally under the Montreal Protocol. The Chilean private sector and government have been looking for alternatives to the use of MB on this fruit, including ionizing radiation. Up to now, irradiation is the most promising alternative since the technological point of view and associated cost, however, to use this technology at commercial level it is necessary to conduct more research and detailed cost studies, as well as, to get the acceptance of the this process for the consumers and the authorities in charge of the plant protection in the importing country.

1. General information.

Chile is a Latin American country of approximately 15 million of inhabitants where the agriculture represents 6.4% of the gross national product (Asociación de Exportadores de Chile, A.G., 1999).

The Chilean fruit production has experienced a huge growth during the nineties, (an average of 12%/year). This has been possible due to a clear free market policy and to a combined effort between the government and the private sector. The major growth and development achieved by the fruit exporting industry in the present decade has transformed this subsector of Chilean agriculture into an important development area of the country's economy. This sphere of activity enjoys many comparative advantages, notably off-season production with respect to the main consumer markets of the Northern Hemisphere, favorable climatic and soil conditions, supply of skilled labor and excellent phytosanitary conditions.

During the season 1997-1998, the exportations of fresh fruit and vegetables, meant an income of approximately US $1.600 millions, being the main product: table grapes, apples, pears, kiwifruit, plums and nectarines. See Figure N°1. The Chilean fruits and vegetables have 62 countries of destination in the world in all continents. The main destinations of these products are USA (36,5%) and Europe (33,7%).

The establishment of a mandatory treatment for the entrance of commodities to certain markets has a great importance in the Chilean international trade. In this respect, a series of species or products require a quarantine treatment, such as the table grapes, kiwifruits, lemons and tomatoes and also other products where a quarantine treatment is presented as an alternative to the inspection (ARCAL, 1999).

The production of table grapes (Vitis spp) for export is very important in the III, IV, V, VI Regions and the Metropolitan Region of Chile. In the last season (1997-1998), Chilean producers exported a quantity of 67.148.052 boxes of grapes, equivalent to 506,188 metric tons and total revenue of US $650 millions. From the total volume of table grapes exported to different markets, approximately 55,5%, (37,3 million boxes) was channeled to the market of USA. (Asociación de Exportadores de Chile, A.G., 1999). See Figure Nº2.

The fumigation of table grapes is mandatory for the Brazilian and USA market due to the presence of a small mite: *Brevipalpus chilensis Baker* (false spider of the vine). The reduced size of *B. chilensis* (<1mm) makes the inspection of the shipments unfeasible. This situation has determined the fumigation with MB as the surest alternative of quarantine treatment.

The Methyl Bromide (MB) is a very effective chemical fumigant, but is being phased-out globally under the Montreal Protocol (an international treaty for the regulation of ozone depleting substances worldwide and under the auspices of the United Nation Environmental Program), because in 1991 it was identified as a destructive substance of the layer of ozone. The Montreal Protocol has a schedule for the phase-out of MB: year 2005 for advanced countries and 2015 for developing countries. Up to now, there is an exception in the prohibition of MB for quarantine and pre-shipment (phytosanitary) but this do not imply this situation will remain in the same way in the future (UNEP, 1995)

MB is the principal method of import quarantine treatment approved in USA. In 1993, the United State Environmental Protection Agency (EPA), under the auspice of the Law of Clean Air (CAA) proposed a total prohibition of consumption and production of this fumigant by 31 December 2000, with no exception (FAO/IAEA, 1997). The USA decided to change its law, which was earlier promulgated to ban the production and consumption of MB by January 2001 to be consistent with the phase-out schedule of the Montreal Protocol. Thus, the production and consumption of MB could still valid until 2005.

The economic impact of a prohibition of MB, especially in the USA, could be of enormous proportions for Chile, since it should not only be considered the possible loss of the more important market for Chilean table grapes, but also the negative repercussions that would exist in the sectors associated to the production and exportations of table grape, such as: factories of containers and packings, transport, chain of cold, manpower and the low use of the infrastructure of ports and airports etc. In fact, the prohibition of entrance for Chilean table grapes to the North American market in 1989, during 2-3 weeks, it meant a loss in income of US $300 millions for the country (Araya, E. 1999).

The present paper, will try to summarize the most important aspects related to the possible use in Chile of the ionizing radiation as a quarantine treatment on table grapes to be exported, and especially in what concerns to the North American market.

2. Technical feasibility of applying ionizing radiation as a quarantine treatment.

2.1 Effect of low doses of irradiation on commercial quality.

Commodity quality is a very important subject related to the feasibility of ionizing radiation as a quarantine treatment, because many fruit suffer some deleterious effects by irradiation, however, the fruit's response depends strongly of factors such as variety, maturity, cultivar, time between harvest and treatment, post treatment handling, packaging, doses and dose rate.

Based on the factors mentioned, by the middle of the eighty, the Chilean Nuclear Energy Commission together with one of the main exporters of fruits of that time, decided to begin the studies on the effect of low dose of ionizing radiation (<1 kGy), in the commercial quality of 19 species of fruits, including different varieties of table grapes, such as Flame seedless, Ribier and Thompson seedless. The parameters considered in this studies were: pH, acidity, soluble solids, loss of weight (%), color and sensorial evaluation (CCHEN, 1990). The results of the studies carried out, indicated that doses <1 kGy did not affect the quality of the most analyzed fruit, under the normal exportations and commercialization conditions of this product in Chile (exception was obtained for cherimoyas and sweet cucumber).

Furthermore, a complementary study was carried out through a research contract with the IAEA (N ° 4456/R1/RB). This study also included table grapes and it checked that an average dose of 1 kGy did not affect the commercial quality of this fruit (Rubio, 1991).

2.2 Effect of low dose of ionizing radiation on *Brevipalpus chilensis*.

At the middle of eighty, the first studies on the effect of irradiation in different stages of development of *Brevipalpus chilensis* were carried out (Sánchez, 1989, Sazo and Gerstle, 1989).

A second study was done during the period 1993-1994 (Jadue et al, 1996), where mites were irradiated on individual excised *Ligustrum japonicum L.* leaves, at different stages: 1-4 and > 4 days old eggs, juveniles and adults. For each treatment 40-50 individuals were used, with 4 replicates. In order to simulate the export conditions, the material of study was stored in a cold chamber by 24 hrs. previous to the irradiation and then it was maintained for 15 days to 0-2°C and 90% of RH. The accumulative hatched (%) of eggs, the mortality of the juvenile and mature stages, as the oviposition and viability of the eggs coming from adults treated with irradiation were studied, during 15-19 days of storage at 25 + / - 1°C and 50-70% of RH. The data were analyzed statistically, using the ANOVA and Duncan's test of multiple ranges. The results are presented in the Figures N ° 3,4,5,6,7 and 8.

It is observed in the Figures N° 3 and 4 that irradiation, followed by cold storage, kill eggs of both stages of development. This effect was accompanied by visual changes of the eggs such as color and dehydration. The eggs without irradiation treatment generally presented a more intensive red color and without signs of dehydration.

In the Figures N° 5 and 6, it can be observed that doses of 1,5 kGy produced 100% of mortality of the juvenile stages and adults since the first day of sampling. The lower dose of 1,0 kGy produces the same effect (100% of mortality) of juvenile stages. All adults were killed by 0,5 and 1,0 kGy on days 17 and 9 of sampling, respectively.

The visual observation of the juvenile stages and adults demonstrated also color changes (paler) and dehydration.

The eggs, produced by the oviposition of a few mature survivors irradiated at 0,5 kGy (4,27% the day 19 of the sampling) did not continue their development. See Figures N° 7 and 8.

In general terms, it can be said that the eggs were more radiosensitive than the juvenile stages and these in turn more sensitive than the adults do. For eggs, juvenile and mature states, dose range of 0,5-1,5; 1,0-1,5 and a dose 1,5 kGy produced 100% of the mortality, respectively; so it could reach the standard for required security based on the concept of the probit 9 (32 survivors in a million). Similar results have been reached by other authors (Sanchez, 1987; Ignatowicz, 1990).

At present, the studies have continued with a combined effort of the Federación para el Desarrollo Frutícola (FDF) (corporation of private right with 53 exporters companies) and the Chilean Nuclear Energy Commission (CCHEN). These new studies have the support of IAEA trough a Coordinated Research Project (research contract N° 10471/R0) and the Ministry of Agriculture of Chile (Agricultural and Cattle Service). These new studies will check the previous results using this time table grapes artificially infested with *Brevipalpus chilensis* and according the recommendations of the investigation protocol proposed by the IAEA (IAEA, 1999)

2.3 Legal aspects

The technical feasibility necessarily includes legal aspects. At present, a group of 41 countries (including USA, Chile and Brazil) have approved the consumption of food irradiation in a great variety of products, and only 14 countries of this group have approved the treatment of fresh fruits for their desinfestation, fruit by fruit base or general base (fruit group) (IAEA, 1998). Although ionizing radiation is recognized as a quarantine treatment, there was no economic incentive in using it, because the wide application of fumigants, however, this situation is changing very fast.

Taking into account that the main destination of Chilean table grapes is USA; it is significant to note that US Food and Drug Administration (FDA) approved in 1986 the irradiation of fresh fruits for its desinfestation or delay of ripening, with maximum dose of 1 kGy. Additionally, the irradiation as a quarantine treatment was recognized for the North American Plant Protection (NAPPO) already in 1989. This decision was imitated later on by Agencies of Agricultural Protection of other countries. However, the United States Department of Agriculture (USDA), through the Animal and Plant Health Inspection Service (APHIS) has issued a specific regulation on irradiation as quarantine treatment only for three fruits from Hawaii against the fruit fly: papayas, litchis and carambola (USDA, 1997). Thus, it has sent approximately 240 tons of fruits from Hawaii to the continent (Hallman, G. 1999). These authorizations are very important for the commercial interchange between Chile and USA, because USDA has taken a decisive step by allowing ionizing radiation as a quarantine treatment of fruits not coming from North America continental land. Before this authorization some special permissions were given by USDA/APHIS to State of Hawaii to treat fruit (papaya, lychee, rambutan, carambola, cherimoya) in small irradiator located in Chicago, Illinois. Additionally, NAPPO has published a guidelines for the use of irradiation as a phytosanitary treatment (NAPPO, 1997).

At international level, it is important to take into account the Agreement adopted during the Uruguay Round of GATT Multilateral Trade Negotiation, because governments which have import regulation stricter than recognized international standards, guidelines and recommendations, may be requested to furnish justifications based on scientific grounds to the World Trade Organizations (WTO). Additionally, international organizations like Codex Alimentarius Commission and advisory groups like International Consultative Group of Food Irradiation (ICGFI, 1994) have established worldwide rules related to food irradiation including its use as quarantine treatment.

At present, other important effort have been made on legal aspects, such as the establishment of a model of regulations in some regions, like Latin America (ICGFI, 1997), Africa and Asia, and a new regional project for Latin America on the harmonization of a regulation to use ionizing radiation as a quarantine treatment for fruits.

3. Economic aspects.

Along with evaluating the technical efficacy and legal aspects, the relative costs of irradiation are central to assessing its potential as a treatment alternative.

At the beginning of the nineties a design of an irradiation commercial plant was studied in order to treat huge volumes of table grapes to be exported from Chile to USA, using a maximum dose of 1 kGy. For this purpose many aspects were taken into account like volume of exportations, seasonal peaks, areas of destiny, flows of trucks, doses to be applied, size of boxes, pallets etc. Chilean engineering and some imported components, including Co-60, were considered to calculate the costs of investment (Galvez and Ramirez, 1990)

From this analysis it was determined that a commercial plant irradiating a volume over 15.000.000 boxes of table grapes for season (6 months) was profitable. The price of the service was US$ 20 cents per box (3,2 to 3,4% of sale price of the product). Similar results were obtained by Forsythe and Evangelou, 1993, based on a number of assumptions. Including the treatment in a port of USA. They obtained a price of US$ 23 cents per box.

At present, the cost of MB in Chile is 18-20 cents US$ per box, however it is expected an increase in MB cost in the future, due to the reductions of its offer in the international market.

During the last years, the Asociación de Exportadores de Chile has been analyzing other alternatives to the use of MB, like the recycle of MB, pulsed UV light (laser), microwaves etc. Up to now there is no technical and economically feasible alternative for the use of MB. Ionizing radiation is the most promising alternative.

4. Acceptance of the process.

Two other aspects are necessary to solve before the commercial application of ionizing radiation as a quarantine treatment to be feasible: i) the approval of the technology for the government regulatory agencies of importing countries based on the prevention of reproduction of adult and not on the mortality and ii) the consumer acceptance.

It is well known that the irradiation with specific dose prevents the reproduction or the development by pests organisms that infest a host commodity, but it may not eliminate or destroy them outright. As a result of this effect a few organisms can survive and this hinders the acceptance of the process for the authorities in charge of the plant protection aspects. The inspectors will have difficulty differentiating between treated and untreated pest.

If the irradiation is a mandatory quarantine treatment in the future, the process must be supported by good agronomic practices and good irradiation practices, including standards dosimetry procedures and the certification of the consignments signed by the responsible quality control official.

Although, there is some scientific knowledge about the effect of irradiation on some insect and mites of quarantine importance for table grapes, it is necessary to carry out more research in order to support a future acceptance of the process.

The consumers acceptance is other important subject to be solved but it has been demonstrated by a great quantity of studies that a good campaign of diffusion and information can be decisive. Up to know, the consumer acceptance of food irradiation in USA seems to be not great problem, the evidence suggests a positive consumer response.

5. Conclusions.

- The phase-out of MB as a quarantine treatment of fresh fruits is a topic of a great importance for the international market of Chilean fruit and especially for table grapes send to the USA.

- Ionizing radiation is being seriously considered in Chile for table grapes, as an alternative to MB.

- There is important data on the effect of low doses of irradiation on the commercial quality of fruits and as a method for disinfestation, specially on *Brevipalpus chilensis*, but additional research and economic evaluation will be required before its use.

- The feasibility of applying this process at commercial level for quarantine purposes on Chilean table grapes depends on several factors, including the acceptance of the technology for the consumers as well as for the government regulatory agencies in importing countries.

- Important advances on legal aspects of ionizing radiation as a quarantine treatment have been achieved during last years at national, regional and international level. These regulations will be an important support for the use of irradiation as a quarantine treatment at commercial scale in the future, including the treatment of Chilean table grapes.

References

Araya, E. Personal communication. 1999.

ARCAL (Acuerdo Regional de Cooperación para la Promoción de la Ciencia y la Tecnología Nucleares en América Latina). 1999. Informe de la Primera Reunión de Coordinadores del proyecto RLA/5/043 (ARCAL XXXIII) " Reglamento cuarentenario armonizado para la irradiación de frutas ", Habana, Cuba. 22 al 26 de Febrero de 1999.

Asociación de Exportadores de Chile. A.G. 1999. Documento Estadístico. Temporada 1997/1998. Santiago. Chile.179 p.

Comisión Chilena de Energía Nuclear (CCHEN). 1990. Efecto de bajas dosis de irradiación en la calidad comercial de frutas y vegetales frescos de exportación. Período 1985-1989. Informe Interno CCHEN.1990. 49 p.

FAO/IAEA. 1997. Irradiation as an alternative to Methyl Bromide Fumigation of Food and Agricultural Commodities. Proceedings of a Consultant`s Meeting. Vienna. Austria, 11-15 August. 1997. 38 p.

Forsythe, K. Jr. and Evangelou, P. 1993. Costs and benefits of irradiation and other selected quarantine treatments for fruit and vegetable imports to the United States of America. In costs benefits aspects of Food Irradiation processing. Proceeding of a Symposium, March 1-5, 1993, Aix-En-Provence, France, p.271-290.

Galvez, L y Ramirez, R..1990. Anteproyecto de una Planta de Irradiación gamma para uva con fines cuarentenarios. Tesis para optar al título de Ingeniero Civil. Universidad de Santiago de Chile. Santiago, Chile. 176 pág.

Hallman, G. Personal communication 1999.

Ignatowicz, S. 1990. Irradiation as a quarantine treatment for agricultural products infested by acarid mites (Acari : Acaridae). In Use of irradiation as a quarantine treatmente of Food and agricultural comodities. Proceedings of the Final Research Co-ordination Meeting, August 27-31, 1990, Kuala Lumpur, Malasya, p. 157-174.

International Consultative Group of Food Irradiation (ICGFI).1994 Irradiation as a quarantine treatment of Fresh Fruits and Vegetables. ICGFI Document N°17. IAEA. Vienna 1994.

International Consultative Group of Food Irradiation (ICGFI).1997. Taller sobre armonización de regulaciones en irradiación de alimentos en América Latina y el Caribe.Lima. Peru. 21-25 Abril 1997. Documento GCIIA N° 25. Viena. Austria. 1997.

International Atomic Energy Agency (IAEA). 1998. Food and Environmental Protection. 1 (2) IAEA. Vienna. Austria. 1998.

International Atomic Energy Agency (IAEA). 1999. The First FAO/IAEA Research - Coordination Meeting (RCM) on Irradiation as a Phytosanitary Treatment of Food and Agricultural Commodities. 29 March- 2 April. Bangkok. Thailand.

Jadue, Y; Vargas, C; Araya, J; Rubio, T. 1996. Preliminary evaluation of ionizing radiation as a quarantine treatmente for the false grape mite, Brevipalpus chilensis.
Zetschrift für Pflazenkrankheitn und Pflanzenschutz. Sacz :11/96. ISSN 0340-8159

NAPPO (North American Plant Protection Organization). 1997 Standards for Phytosanitary Measures.. Guidelines for the use of Irradiation as a Phytosanitary Treatment NAPPO Newsletter, Vol 17, N ° 3 (July 1997).

Rubio,T. Effect of low doses (< 1 kGy) in the commercial quality of fresh fruits and vegetables for export. Food Irradiation Newsleter. 15 (1), May 1991. p16-17.

Sanchez, F. 1987. Irradiation treatment to sterilize Naupactus xanthographus and Brevipalpus chilensis females in table grapes. Food Irradiation Newsletter 13 (1) p.37- 38.

Sazo, R y Gerstle,C. 1989. Efecto de las radiaciones de Cesio 137 sobre la mortalidad de Naupactus xanthographus Germar (Coleóptera : Curculionidae) y Brevipalpus chilensis Baker (Acari : Tenuipalpidae). Ciencia e Investigación Agraria 16 (12) p. 65-67.

Sepúlveda, M. y Villanueva, L. 1996. Tratamicnto de las emisiones de Bromuro de Metilo en cámaras de fumigación . Universidad de Chile – Ministerio de Agricultura.. 33 5 p.

United Nations Environment Programme (UNEP). 1995. Montreal Protocol on Substances that Deplete the Ozone Layer. 1994 Report of the Methil Bromide Technical Options Committie. 1995. 288 p.

United States Department of Agriculture (USDA). 1997. Papaya, Carambola and Litchi . Final Rule. 7CFR. Parts 300 and 318. Federal Register. 62. (132.) pp 36966-36967 (July 10 . 1997)

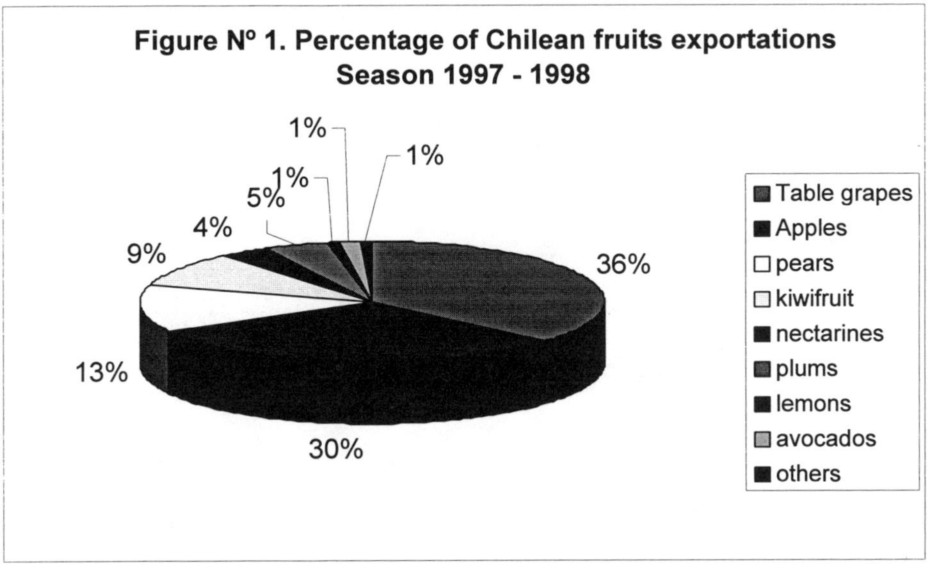
Figure Nº 1. Percentage of Chilean fruits exportations
Season 1997 - 1998
1%
1%
1%
5%
4%
9%
36%
13%
30%
Table grapes
Apples
pears
kiwifruit
nectarines
plums
lemons
avocados
others

Figure Nº 2. Destination of Chilean Tables Grapes
Season 1997 - 1998

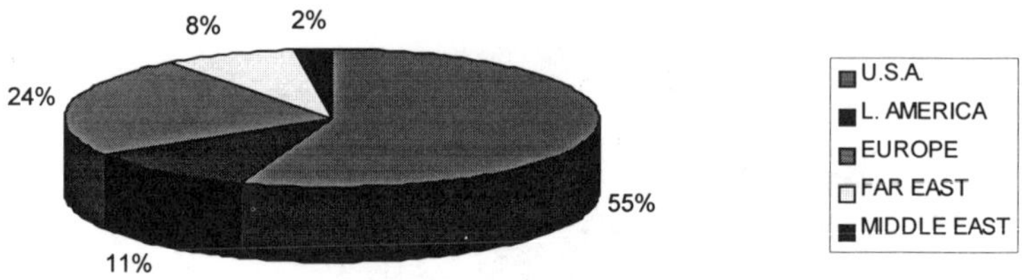

Figure Nº 3: Accumulative hatch (%) of irradiated and non irradiated 1 - 4 days old eggs of Brevipalpus chilensis

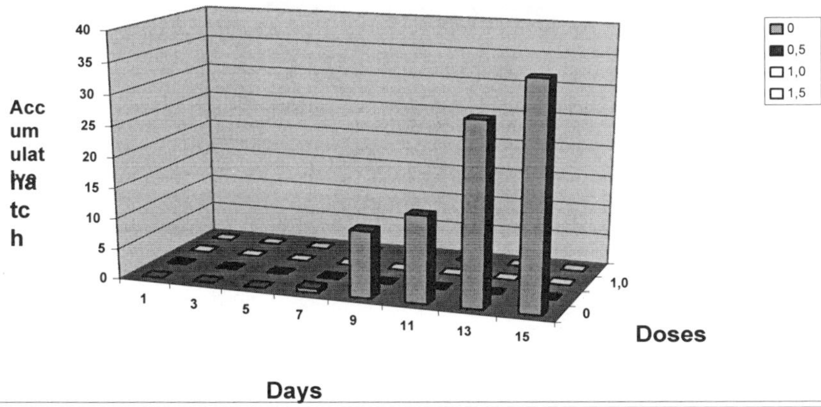

Figure Nº 4. Accumulative hatch (%) of more than 4 days old eggs of Brevipalpus chilensis

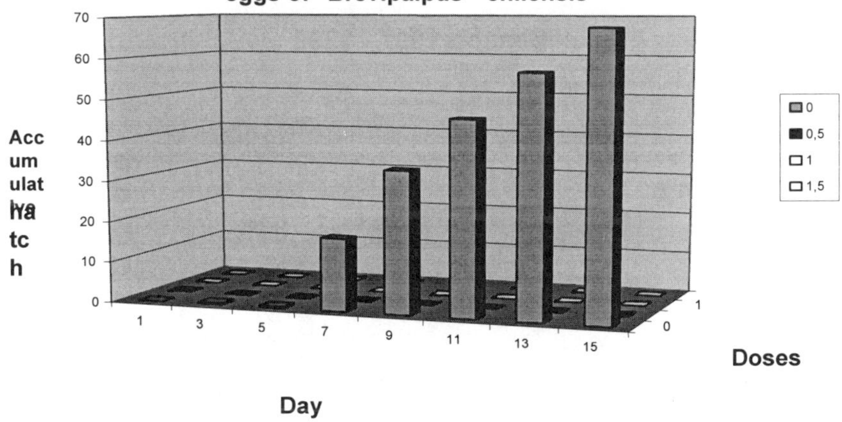

Figure Nº 5. Accumulative mortality (%) of irradiated and non irradiated juveniles of Brevipalpus chilensis

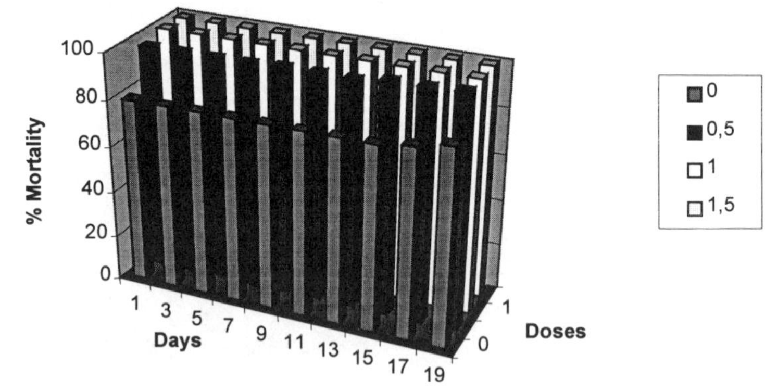

Figure Nº 6. Accumulative mortality (%) of irradiated and non irradiated adults of Brevipalpus chilensis

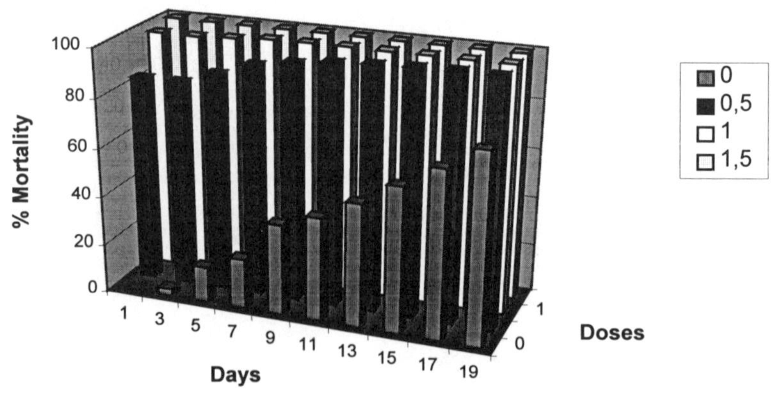

Figure Nº 7. Accumulative oviposition (%) of irradiated at 0,5 kGy and non irradiated adults Brevipalpus chilensis

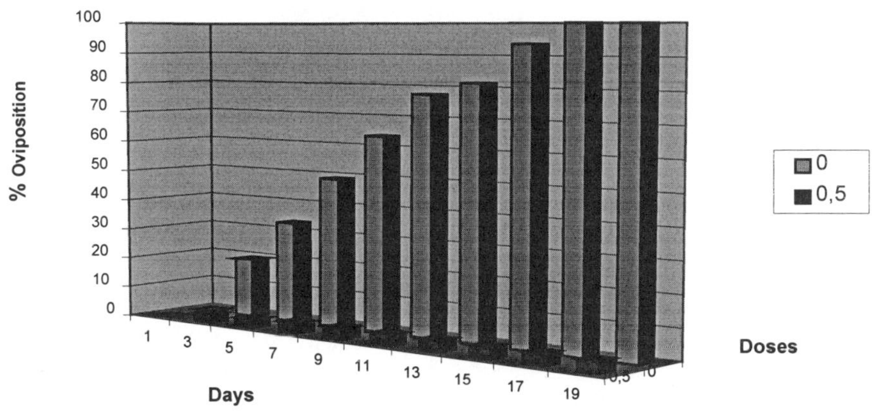

Figure Nº 8. Accumulative eggs hatch(%) of irradiated at 0,5 kGy and non irradiated of Brevipalpus chilensis

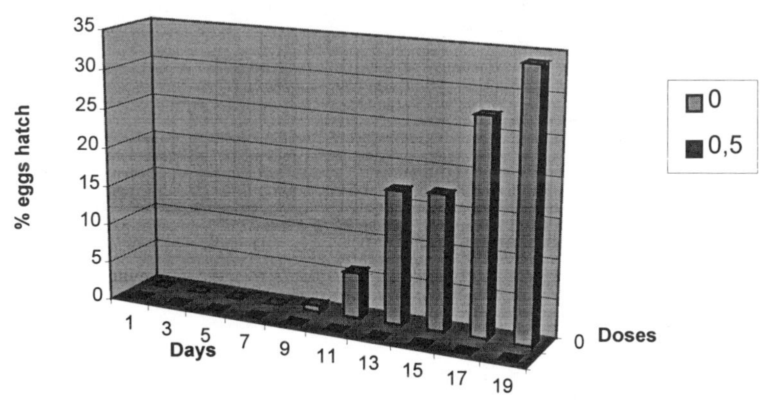

Prospects of Radiation Sterilization of Shelf-Stable Food

I. DE BRUYN
Atomic Energy Corporation of SA
P.O. Box 582
Pretoria, 0001, South Africa

ABSTRACT

The basic procedure developed by the U.S. Army to produce irradiated shelf-stable foods was adapted to develop a number of meat dishes with gravy. The meat dishes are of a high quality and have a good taste and texture profile. Starches and vegetables could not successfully be irradiated at high dose. Between 1989 and 1995 approximately1.8 million portions of shelf-stable meat products were sold to the South African Defence Forces (SADF), however, due to the restructuring in the South African government and therefore the SADF no contracts are currently in existence. During 1992 approval of the South African Department of Health was obtained to sell to outdoor enthusiasts through selected hiking/outdoor shops that act as agents. Shelf stable meats have been used by the 1st South African Mount Everest Expedition, Cape to Rio international yachtsmen and outdoor adventures such as hiking, river rafting , pony trekking etc. The product is also used by exploration teams and as an emergency ration by mining companies.

Market surveys following exposure of consumers to audio-visual information and tasting of the shelf-stable food indicated that 76% would buy the product and only 5% said they would not buy, a trend similar to that reported by researchers in the United States. The details of the process parameters in respect of irradiation dose, packaging , conditions during irradiation and various high quality shelf-stable products developed are discussed.

1 INTRODUCTION

During the 1950's to end 1970's the United States Army developed the basic methodology to produce shelf stable irradiated meat, seafood and poultry products. These products are normally packed without gravy, sauce or brine, as liquid is not required to sterilize the product as in the canning process. This leads to the distinctive "dried cooked" taste normally associated with roasts opposed to the casserole taste usually associated with tinned meats.

The Biogam group at the Atomic Energy Corporation of South Africa is currently producing shelf stable irradiated meats on a commercial basis. The meats are cooked, chilled, portioned, vacuum packed and irradiated to the required minimum dose of 45 kGy at a temperature of between -20 and -40 °C to ensure absolute sterility even under tropical conditions. The product is packaged in a high quality four layer laminate pouch and will therefore not rust or burst even under adverse weather conditions and can be guaranteed for more than two years as long as the integrity of the packaging is maintained.

2 HISTORICAL BACKGROUND

Researchers have been experimenting with the irradiation of foodstuffs for more than 100 years. In 1943 researchers at MIT published a paper on "Effect of X-ray irradiation on the bacterial count of ground meat". During the early 1950's isotope sources became available and research started in earnest. These studies gained greater impetus with the "Atoms for Peace Program" which President Eisenhower proposed to the UN in 1953.

The groundbreaking research into irradiation of food at high doses was carried out in the USA. Between 1953 and 1962 studies were undertaken by the Army's Laboratories at the Quartermaster Food and Container Institute in Chicago, Illinois, while between 1962 to 1980 research was continued at US Army Natick R&D Command in Natick, Massachusetts [1], [2], [3]. In 1963 FDA clearance was obtained to irradiate bacon at high dose. The irradiation of packaging materials to be used as food contact layer was approved between 1964 and 1967 [1]. Unfortunately during 1958 the Food Additives Amendment to the Federal Food, Drug and Cosmetic Act was tabled whereby irradiation was regarded as an additive. The Act specified that food had been adulterated if it had been irradiated intentionally. Such foods can therefore not be sold unless the food is irradiated under prescribed conditions that have been proven to be safe [4].

In 1966 the US Army petitioned the FDA to approve irradiated ham based on the approval of bacon. This petition was finally turned down in 1968 and in the same year the clearance for bacon was revoked [1], [2]. This lead to the reduction of most high dose research programs throughout the world [2]. Following the FDA's requirements the protocol for animal feeding studies to determine the wholesomeness of irradiated food was changed considerably [1]. The results of these studies were used as a major source of data for the WHO appointed Study Group which published it's conclusions in a Press Release in September 1997 as: "foods treated with doses greater than 10 kGy can be considered safe and nutritionally adequate when produced under established Good Manufacturing Practice." [5].

One of the first commercial applications for shelf stable irradiated food was in 1977 when the Fred Hutchinson Memorial Hospital in Seattle placed an order for pre-cooked irradiation sterilized hospital meals for use by bone marrow transplant patients [6]. Unfortunately in 1988 the Fred Hutchinson Memorial Hospital stopped using radiation sterilized meals, as the Cobalt source that had been used to sterilize the products became too weak [7]. In 1995 the FDA approved the irradiation of frozen meals at a minimum dose of 44 kGy for use by NASA astronauts [8].

3 PROCESS PARAMETERS

3.1 INTRODUCTION

The basic procedure used to irradiate shelf stable products was developed at Natick and can be summarised as follows [7]:

FIGURE 1: Irradiation process

Cook to attain internal temp of 73 to 77°C
(Inactivation of enzymes)

Slice or portion as required

Vacuum pack
(Multi-layer aluminium foil barrier pouch)

Freeze to -40°C
(Cryogenic or mechanical)

Irradiate to minimum dose of 25 to 45 kGy[a]
(Temperature between –40 and -20°C)

[a] The minimum dose is determined by presence or absence of nitrites as described in section 3.2

3.2 MICROBIOLOGICAL, CHEMICAL AND TEMPERATURE PARAMETERS

The dose required to produce irradiation sterilized products has to be sufficient to kill off 12 log cycles of *Clostridium botulinum,* a highly toxic, radiation resistant anaerobic organism [10]. Some of the results that were obtained at Natick to determine the minimum irradiation sterilizing dose required to ensure absolute sterility of the products are listed in Table 1. It must be noted that the nitrite containing products like corned beef, bacon and pork sausage have a lower 12D and therefore lower minimum sterilization dose due to the anti-botulinal effect of nitrite.

TABLE 1. Minimum irradiation sterilising (12D) doses in kGy [9]

Food	Irrad. Temp	Method of establishing 12D dose[a]	
	(°C)	Extreme value[b]	Spearman-Karber[c]
Beef	-30 ± 10	41.2	43.3
Chicken	-30 ± 10	42.7	44.3
Ham	-30 ± 10	31.4	38.1
Pork	-30 ± 10	43.7	39.2
Codfish cake	-30 ± 10	31.7	32.4
Corned beef	-30 ± 10	26.9	24.4
Pork sausage	-30 ± 10	25.5	26.5
Bacon	5 to 25	-	25.2

Source: RB Rowley, Natick labs

[a] Based on recoverable botulinal cells and an assumed one most resistant strain/can

[b] Based on an assumed exponential spore death rate with an initial shoulder
[c] Based on an assumed exponential spore death without an initial shoulder

During the planning of an inoculation pack study to determine the minimum irradiation dose, the envisaged processing parameters should be adhered to strictly as changes in temperature during irradiation [11] as well as the chemical formulation of the meat product [12], [13] have a significant effect on the 12D value.

3.3 PACKAGING REQUIREMENTS

Originally tinplate cans were used for the packaging of shelf stable foods, however with the development of flexible pouches this is now the preferred material. The requirements for the packaging material were [14]

- must be easily heat-sealed
- must withstand the processing at -40°C without delaminating, cracking, or losing seal strength
- must withstand shipment and handling hazards
- must provide a total oxygen and moisture barrier
- must be inert to the food in the pouch
- must protect the food from microbial or other contamination

Because of potential changes in the material caused by irradiation, the safety of any material has to be tested before use. Migration of chemicals in the packaging can be affected by irradiation and harmful volatile decomposition compounds can potentially migrate directly into the food. FDA therefore requires that food irradiation packaging materials must conform to regulations (21 CFR 179.45) and be based on appropriate testing [4]. The material used by Natick is a three-layer composite consisting of: Nylon 6 film on the outside, aluminium foil in the middle and polyethylene terephthalate-polyethylene film as the food contact layer [14].

4 ADVANTAGES AND QUALITY OF IRRADIATED FOODS

Irradiation leads to a high quality product, which cannot easily be distinguished from a similarly processed equivalent. Because of the process the products can be served in bulk as whole beef or pork roasts, or in individually portioned servings. These products have a typically "dry cooked" flavour which cannot be produced by any other sterilization process. The process reduces or eliminates the need for chemical preservatives. The product normally has a higher nutritional value than heat sterilized food because there is less destruction of vitamins and essential amino acids [9],[15].

Irradiation tenderises the meat by degrading connective tissues but meat must be trimmed of bone, excess fat and gristle. Irradiated meats are slightly drier than meat products produced by other processes, but this can be solved by the addition of sodium chloride and phosphates prior to cooking which ensures a tender, juicy end-product [3].

In the canning process the term "commercial sterility" is used which means that canned food is sterile at ambient temperatures of 20 to 25°C. However under tropical conditions certain thermophilic organisms such as *Bacillus stearothermophilus* can still grow and cause spoilage of the food. Irradiated food, however, is sterile under all temperature conditions as the process is designed to kill all spores of *Clostridium botulinum* which is the most radiation resistant bacteria to survive the initial cooking process of the meat. Shelf stable irradiated meat products that are produced in flexible pouches can be used in temperature conditions ranging from -60 to +60°C and are therefore ideal to use from the top of Mount Everest to the tropics.

5 THE SOUTH AFRICAN EXPERIENCE

5.1 INTRODUCTION

During the late 1970's the South African Defence Force requested the Atomic Energy Corporation of South Africa (AEC) to develop shelf stable meat products for use under special conditions. The radiation sterilization techniques developed by the USA Defence Force laboratories at Natick [7] were therefore optimised for South African conditions.

In all cases meat with low internal muscle fat is used and bone, gristle and excess fat is trimmed off the meat before cooking to an internal temperature of 73 to 77°C. The cooking process inactivates the proteolytic enzymes which are only slightly affected by the irradiation. Vacuum packaging is done in a four laminate flexible pouch consisting of Nylon/Foil/ Polyester/LLDPE. The freezing is mechanical and the product freezes down to -40°C in about 8h. During irradiation the product is maintained at deep frozen temperatures by placing dry ice on top of the product in the irradiation tote box.

The dose used throughout is a minimum of 45 kGy as that dose was found to be equivalent to a 12D dose for the "worst case" product i.e. Beef stew without any preservatives or phosphates. Inoculation pack studies were carried out using *Clostridium sporogenes* to establish the 12D dose [16]. Nitrite containing products like bacon can be irradiated at a minimum of 26 kGy, which provides 12D safety due to the added anti-botulinal effects of the nitrite [17]. However, for practical reasons it was decided not to irradiate different products at different minimum doses.

The South African Bureau of Standards (SABS) requires that the total production be incubated at 37°C for 10 days and not just a percentage as is done in the canning industry. They are slightly wary of this "new" technology and are strictly monitoring the production. After completion of the incubation, sampling is done by the SABS and microbiological and sensory evaluation of the product is carried out. A Certificate of Compliance is then issued before the product may be sold.

5.2 MEAT PRODUCTS

A number of meat dishes were successfully developed: Roast beef with gravy, Roast chicken with sauce, Beef steak with tomato, Bacon, Beef curry, Beef stroganoff, Bobotie, Roast pork with gravy, Chicken curry, Chicken and tomato, Pork country sausage and Savoury mince with mushrooms. The meat products are of high quality and have a good taste and texture profile, although optimisation of some of the sauces is still required [7].

Studies have shown that starch based thickeners used with the meat products are seriously affected by the freezing [18]. A pea protein based thickener has been used to create a stable sauce, but the protein imparts an unacceptable taste and it can therefore not be used at present. This work is still underway and final results are therefore not available.

5.3 VEGETABLES

Experimentation was carried out with various vegetables. Initial studies were carried out by cooking the vegetables, which were then frozen and irradiated. It was found that the products became very soft due to the breakdown of the cellular structure during processing. Subsequently the products were only blanched and then packed, frozen and irradiated. After irradiation the products were initially marginally acceptable but after a three month period the product again became very soft and unacceptable from a sensorial point of view.

In summary it is possible to produce some shelf stable vegetables but on the whole they are not very appetising in appearance, taste and texture and are fairly expensive due to the process. Canned vegetables of equivalent or better quality are available at roughly a third of the price. Vegetable dishes could possibly be investigated since the price is not such a problem with value added products.

5.4 STARCHES

Initial work with starch products like rice, noodles, potatoes and various dried beans showed that they became very soft after processing and it was therefore decided to do additional studies using only partially cooked products. However par-boiling did not solve the problems either as the products were still unacceptable after processing due to the very soft texture.

With the exception of kidney beans (which contain a high percentage of protein) none of the starch items tested were acceptable because of the damage to the starch compounds by the freezing and irradiation and the subsequent moisture loss. All of the products were edible but not palatable and the cost of the process and packaging is again a major stumbling block in contrast with the meat where the raw product is already a much higher priced commodity and can therefore absorb the process costs.

Details of the South African research into shelf stable food products can be found in Appendix B of IAEA-TECDOC-843 [7].

6 SALES OF IRRADIATED MEAT PRODUCTS IN SOUTH AFRICA

6.1 MILITARY SALES

During 1989 general clearance was obtained from the South African Department of Health to supply shelf stable irradiated food to the South African Defence Force (SADF). Between 1989 and 1995 approximately 1.8 million portions were sold to the SADF. Due to the restructuring in the South African Government and therefore the SADF no contracts are currently in existence. The SADF has absorbed members from the three different freedom fighting groups and it's numbers have therefore increased from approx. 60 000 to approx. 120

000 troops. This has gone hand in hand with massive cuts in the defence budget, which allows for few "luxuries" such as irradiated shelf stable foods.

General George Meiring, previous Chief of SA National Defence Force, has written a letter in which he confirms the military use of the food. He states: "Many of our special forces used them (irradiated meat products) for long periods as their sole protein intake, and not a single negative health report is on record. We have found the meat packs to be consistently of the highest quality and with their guaranteed shelf life (without refrigeration) of 3 years, they were invaluable for military and emergency operations." [19]

6.2 OUTDOOR SALES

During 1992 AEC obtained special approval from the South African Department of Health to sell to outdoor enthusiasts subject to certain conditions. Prior to this clearance, outdoor enthusiasts could only buy directly from the AEC. Sales are allowed through selected hiking/outdoor shops that act as agents. Sales to corporate clients are also allowed by obtaining approval from the Department of Health to sell to that specific client as long as it is used in-house and not resold.

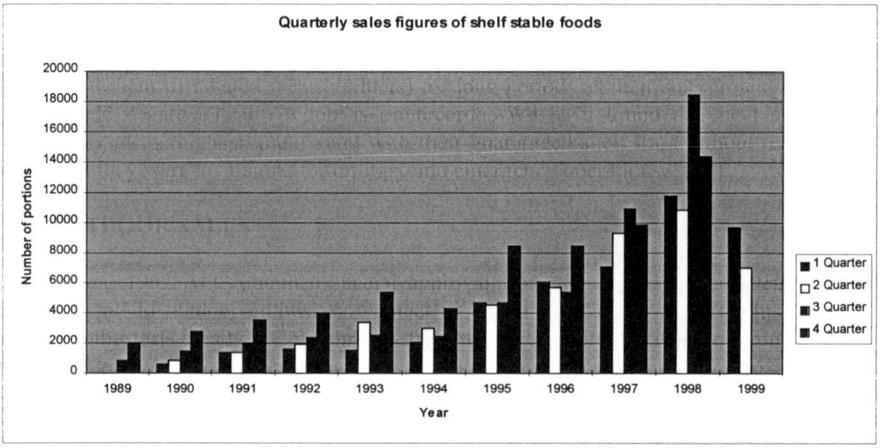

FIGURE 2: Sales of shelf-stable meat products to outdoor enthusiasts in South Africa

The recent decline in sales has been caused by the fact that the price at which the food is sold to consumers has recently been almost doubled as Biogam has now reached the stage where the product can no longer be sold at market penetration prices, but full cost price has to be charged. It is hoped that this trend will be reversed once the consumers become used to the new prices.

Shelf stable meats have been used by:

- 1st South African Mount Everest expedition
- Cape to Rio international yachtsmen
- Hiking and outdoor enthusiasts in South Africa
- 4 x 4 Clubs and Safari's in South Africa and the rest of Africa
- Other outdoor adventures such as river rafting, pony trekking etc.
- Mining exploration teams

Typical comments by some of the users:
"I like it, it can be compacted very small after use and the packaging can be used as a plate in itself... As a quick and easy food, THE BEST I HAVE FOUND YET." - Charlotte du Toit, Camping for Africa (A hiking shop owner), Johannesburg. South Africa.

"Such a wonderful application in difficult cooking situations - yachting especially. Open cooking is dangerous aboard a yacht - this solves that problem. Great "KITCHENEERING!". This is my second irradiated Cape to Rio!" Robin Stamper, Cape to Rio Yachtsman.

7 CONSUMER RESPONSE

A commercial firm of market analysts conducted a marketing survey on the attitude of consumers towards irradiated shelf stable foods. The people attending the session were selected at random from the general population. When asked initially whether they would buy irradiated foods, 15% said they were likely to and 65% said that they were unlikely to.
FIGURE 3: Consumer response to information and tasting of irradiated foods

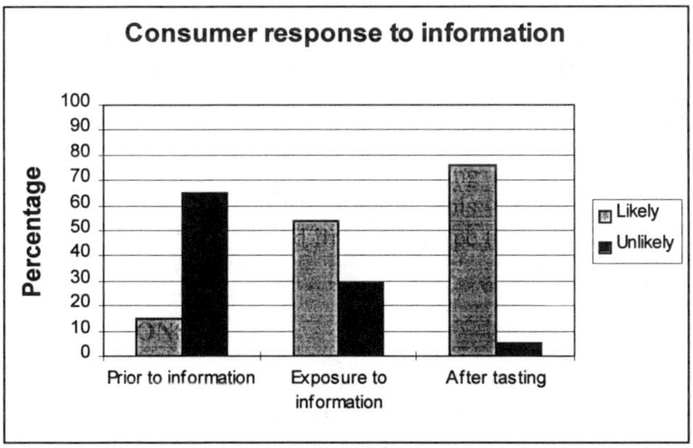

After the consumers were exposed to visual information, 54% said they were likely to buy the food and 29% said they were unlikely to. After audio-visual information and tasting of the food, 76% said they would likely buy the product and 5% said they would probably not buy the food. These results confirm those obtained by various researchers in the United States [20], [21], [22].

8 PROSPECTS FOR COMMERCIAL APPLICATION OF RADIATION STERILIZED SHELF STABLE FOOD

Currently South Africa is the only country that has approved the general sale of irradiated shelf stable foods. In the USA limited approval is available for NASA to use the product. A FDA petition is being drawn up to request approval for sales of shelf stable irradiated foods to the general public. It is expected that this petition will only be granted in 18 months time. The recent approval of the red meat petition by FDA has led to great expectations of large volumes of sales of low dose irradiated foods in the States. This could possibly also affect the future sales of shelf stable irradiated foods.

World Health Organisation has recently (September 1999) published it's final report, two years after the September 1997 conclusions of the Study Group that the process is completely safe. In Europe the prospects of obtaining approval in the near future are very slight due to the requirements set by the Scientific Council of the European Union. The Codex Alimentarius Commission has decided to initiate the formal procedure to amend the Codex General Standard for Irradiated Foods to take the high dose applications into consideration. This amendment however is not expected within the next two years.

9 CONCLUSIONS

High quality shelf stable meat products can be produced using high dose irradiation. The process parameters have been well defined and coupled to Good Manufacturing Practices lead to safe, healthy products. These products are well received by the general public without undue negative reaction to the irradiation process.

Only once all the international approvals have been finalised can the commercial application of this technology become a reality. In South Africa the outdoor market is fairly limited in size with sales of about 22 000 portions expected for the current year. Sales of about 250 000 portions per year are required to make the operation financially viable due to the high overhead costs of the process.

10 REFERENCES

[1] BRYNJOLFSSON, A., The national food irradiation program conducted by the department of the Army. J. Food Process. Preserv. (1979) **3** 125 - 138.

[2] BRYNJOLFSSON, A., Food irradiation in the United States. 26th European Meeting of Meat Research Workers. Colorado Springs, August 31 to September 5, (1980) 172 - 177.

[3] JOSEPHSON, E.S., Radappertisation of meat, poultry, finfish, shellfish and special diets. In: "Preservation of Food by Ionising Radiation". (Ed. ES Josephson and MS Peterson) Vol 3. CRC Press, Boca Raton. 1983. 231 - 251.

[4] PAULI, G.H. & TARANTINO, L.M., FDA Regulatory Aspects of Food Irradiation. J. Food Prot. (1995) **58(2)** 209 - 212.

[5] ANONYMOUS., Food Irradiation - Sky's the limit. WHO Press Release WHO/68. 19 September 1997

[6] WELT, M.A., First commercial irradiation of foods in the United States. 26th European Meeting of Meat Research Workers. Colorado Springs, August 31 to September 5, (1980) 190 - 192.

[7] ANONYMOUS., Shelf-stable foods through irradiation processing. IAEA-TECDOC-843, November 1995.

[8] ANONYMOUS., Irradiation in the production, processing and handling of food. 21 CFR Part 170. [Docket nos. 89F-0011 and 93F-0384]. Federal Register. (1995) 60(45) Wednesday, March 8. 1266.

[9] WIERBICKI, E., Technological feasibility of preserving meat, poultry and fish products by using a combination of conventional additives, mild heat treatment and irradiation. In: Combination processes in food irradiation. Proceedings of international symposium on combination processes in food irradiation. Colombo, 24 – 28 November 1980. (1981) 183 - 203. Vienna: IAEA press.

[10] ANONYMOUS., Review of Data on High Dose (10 - 70 kGy) Irradiation of Food. Report of a Consultation, Karlsruhe, 29 August - 2 September 1994. (WHO/FNU/FOS/95.10). (1995) WHO. Geneva

[11] GRECZ, N., WALKER, A.A., ANELLIS, A & BERKOWITZ, D., Effect of irradiation temperature in the range -196 to 95C on the resistance of spores of *Clostridium botulinum* 33A in cooked beef, Can. J. Microbiol. (1971) **17** 135 - 142.

[12] URBAIN, W.M., Food Irradiation. (1986) London: Academic Press.

13] KRABBENHOFT, K.L., CORLETT, D.A., ANDERSON, A.W. & ELLIKER, P.R., Chemical sensitisation of Clostridium botulinum spores to radiation in meat. Appl. Microbiol. (1964) **12** 424 - 427.

[14] KILLORAN, J.J., Packaging Irradiated food. In: "Preservation of Food by Ionising Radiation". (Ed. ES Josephson and MS Peterson) Vol 2. CRC Press, Boca Raton. **1983**. 317-326.

[15] WIERBICKI. E., Technology of irradiation preserved meats. 26th European Meeting of Meat Research Workers. Colorado Springs, August 31 to September 5, (1980) 194 - 197.

[16] DE BRUYN I.N. & DE VILLIERS B.E., Commercial application of high dose irradiation to produce shelf stable meat products. 1. Experimental sterilization dose required for shelf stable beef casserole. Presented at the first Research Co-ordination Meeting on "Production of safe shelf-stable and ready-to-eat foods through high dose irradiation processing" Belfast, UK. 9 to 13 September 1996

[17] ANELLIS, A., GREZC, N., HUBER, D.A., BERKOWITZ, D., SCHNEIDER, M.D. & SIMON, M., Radiation sterilisation of bacon for military feeding. Appl. Microbiol. (1965) **13** 37 - 42.

[18] ERASMUS. C., The stability of irradiation-sterilized starch and protein based sauces. MSc Thesis. (1996) University of Pretoria. Pretoria.

[19] MEIRING, G., Chief of SA National Defence Force. Personal comunication (1997).

[20] BRUHN, C.M., Consumer attitudes and market response to irradiated food. J. Food Prot. (1995) **58(2)** 175 - 181.

[21] BRUHN, C.M., Strategies for communicating the facts on food irradiation to consumers. J. Food Prot. (1995) **58(1)** 213 - 216.

[22] POHLMAN, A.J., WOOD O.B. & MASON, A.C. Influence of audiovisuals and food samples on consumer acceptance of food irradiation. Food Technol. (1994) **48(12)** 46, 48, 49.